COMMUNIST CHINESE AIR POWER

COMMUNIST CHINESE AIR POWER

RICHARD M. BUESCHEL

"We will have not only a powerful army but also a powerful air force and a powerful navy."

From the opening address by Mao Tse-tung at the First Plenary Session of the Chinese People's Political Consultative Conference, September 21, 1949

FREDERICK A. PRAEGER, *Publishers*

New York • Washington • London

FREDERICK A. PRAEGER, PUBLISHERS
111 Fourth Avenue, New York, N.Y. 10003, U.S.A.
5, Cromwell Place, London, S.W.7, England

Published in the United States of America in 1968
by Frederick A. Praeger, Inc., Publishers

Library of Congress Catalog Card Number: 68-11317

Printed in the United States of America

Dedicated to
IRISH

ACKNOWLEDGMENTS

My thanks first go to Al Lewis, editor of *Air Progress* magazine, for starting me on this project. All he wanted was a story in ninety days, never realizing that he would have to wait more than two years and provide constant encouragement all along the way. Next, my thanks go to fellow members of the American Aviation Historical Society who were so helpful in digging up leads and tracking down data and pictures, and then to Gilman Park, assistant to the president of Frederick A. Praeger, whose interest in airplanes led to the aircraft descriptions that have greatly enhanced the manuscript.

So many people helped with this effort that it is difficult to thank them all. All have my appreciation. Particular appreciation goes to Lieutenant Colonel C. V. Glines for putting the Defense Department's complete aviation files on Communist China at my disposal; Royal Frey, chief of the Research Division, Air Force Museum, for a wealth of Korean War material; Alma Wade, of New Orleans, secretary to the Jouett Mission to China, for historical notes and photographs on the early days of the PLAAF; William H. Gregory, editor of *Aviation Week and Space Technology*, for help in tracking down leads and data; Edgar Meos, of Tartu, in the Soviet

Union, for original data on Russian aid to the Chinese Communists; Tom Pilkington for his firsthand historical contributions; four unnamed friends in Hong Kong and Taipei for their important help and insight; *The New York Times* for its superior coverage of the continuing story of Communist China; the Chicago Public Library, with specific thanks to Mrs. O'Keefe; the Defense, Air Force, Navy, and State Department research staffs in the National Archives, in Washington; and the authors and publishers of the many books and periodicals consulted in the preparation of this work, several of which are listed in the Selected Bibliography.

Additional thanks for assistance in assembling a unique collection of photographs goes to Chris Beilstein; Peter M. Bowers; David W. Lucabaugh; Peter Keating; Denys Voaden; the Central News Agency of the Republic of China, Taiwan; Mannosuke Toda, of KokuFan; Eiichiro Sekigawa, of Aireview; Lou Tepke, of UPI in Chicago; Wide World Photos, in New York and Chicago; and Milton Wolff, of Eastfoto, in New York.

Finally, my thanks go to my friends and neighbors in Mt. Prospect, Illinois, who helped me with this book. I am particularly grateful to Kenneth Gong for his invaluable translations, Edith Freund for performing what seemed a never ending typing task, and the Chicago and Northwestern Railway for time, as a commuter, to think and write.

R.M.B.

CONTENTS

Acknowledgments vii

PART ONE. COMMUNIST CHINESE AIR POWER 1

I. *Communist China Finds Wings, 1923–53* 3

II. *The Tiger Grows Strong, 1954–64* 31

III. *Self-reliance and the Future, 1965–70* 61

PART TWO. THE AIRCRAFT OF COMMUNIST CHINA 105

Selected Bibliography 217

Index 229

A section of photographs follows page 118.

PART ONE

COMMUNIST CHINESE AIR POWER

CHAPTER 1

COMMUNIST CHINA FINDS WINGS 1923–53

They hide in the clouds along the Tropic of Cancer, show themselves fitfully over the Russian border, and once in a while come close enough to American aircraft in the Far East for identification. They're spotted on the ground visiting Hanoi and sometimes dart out aggressively over the Gulf of Tonkin from the island of Hainan. On rare occasions, they even show up over the Formosa Strait. No question about it, they're there—the aircraft of Communist China, enigmatic representatives of one of the largest air forces in the world, and possibly the tiredest.

It is more than a little alarming to realize how little Americans know about the People's Liberation Army Air Force (PLAAF) of the People's Republic of China (PRC). (Even the name of this air arm, in typical Communist fashion, is almost too long to remember. In U.S. military jargon, it is known as the CHICOMAF.) Yet we consider Communist

China a potential military threat today, and we will increasingly be faced with aerial confrontations with the PLAAF, the only air force in the Far East that comes anywhere near parity in numbers with our own.

The first direct confrontation of the air forces of Communist China with those of the non-Communist world took place in the skies over Korea in late November, 1950. The Chinese had just crossed the Yalu River into North Korea, and Communist aircraft appeared over the rapidly changing front. It seemed that a modern jet-equipped air force had been created overnight out of whole cloth. Once again the myth of an inherent Oriental inferiority in the air was exploded. Victory in the air was dependent on skill, training, and aircraft superiority, a combination largely—but not exclusively—enjoyed by American pilots over Korea. It was obvious that the Chinese Communists were often capable pilots, in spite of the 14:1 "kill" ratio maintained by U.N. forces during the two and a half years of air combat north of the 38th parallel.

Some day American pilots may again face this force in numbers. In recent years, U.S. forces have lost a number of Navy and USAF aircraft to Chinese fighters and in turn have downed PLAAF aircraft. It is important to know what we face, how formidable Communist China is in the air, and something of the background of the PLAAF.

When the Chinese Communist Party was formed in Shanghai in 1921 as the Kunchangtang (KCT), with a young student named Mao Tse-tung one of the twelve founding members, there was little thought of forming a military force, much less an air arm. The conditions in China at the time were chaotic, with numerous war lords exerting their influence over vast territories by means of their personal armies. It was soon apparent that only a strong political base backed by a military force would provide the means of controlling portions of the Chinese mainland and ensuring a degree of security. Thus, in 1923, under instructions from Moscow, the Chinese Communists joined forces with the Kuomintang (KMT), led by Dr. Sun Yat-sen and his able assistant Chiang Kai-shek. Coincident with the union of the KCT and the KMT, in September, 1923, the Soviets sent a military adviser to Dr. Sun's headquarters in southern China, where the Russian, Mikhail Borodin, was given a free hand by Dr. Sun to analyze and reorganize the military forces of the KMT. It was Borodin's task to convert the KMT movement into a militant organization of revolutionaries who could control the many political entities of China. His analysis of the situation and of the personalities involved led to his recommendation to Moscow that the KMT be supplied with arms and equipment to aid a military takeover. Even the untimely death of Dr. Sun Yat-sen, in March, 1925, did not alter the plan.

The first Chinese Communist pilots were trained more than forty years ago, when Soviet military aid and Russian advisers began to reach Chiang Kai-shek's National Revolutionary Army of China. Although Russia was faced with problems ensuing from her own civil war and aircraft were in short supply, Stalin thought the chance of securing a communized China worth the risk, and aid was arranged. The Aviation Bureau of the KMT was set up in mid-November, 1925, under the titular direction of Wang Ching-wei, chairman of the KMT Political Council and Chiang Kai-shek's second in command. Actual direction came from a Soviet adviser identified as Comrade Rogachev, Russian chief of the KMT general staff, and by December the Aviation Bureau, consisting of a flying school and an army air squadron, was completely in Russian hands. Russian-built Aviakhim U-1 primary trainers, copies of the famous British Avro 504K; Aviakhim R 1.M-5 reconnaissance bombers. duplicates of the wartime De Havilland DH-9A with Russian-built Liberty engines; and Russian Junkers-Fili F.13 transports were shipped to southern China, accompanied by Soviet instructors and mechanics.

When the first U-1 Avrushka trainers arrived in the spring

of 1926, the KMT knew it possessed a decisive weapon at last. Flight training in the Russian trainers, establishing a cadre of Kuomintang and Communist pilots who remained with their respective forces for many years, began immediately, and the Chinese were quick to catch on to the potential of tactical air power in support of their armies. Friction between the Chinese pilots and their Russian commanders, many of whom could not speak Chinese, led to disagreements at the KMT general-staff level, and the Russians were returned to adviser status, never again to regain full control of the KMT military forces. Command of the Aviation Bureau was taken over by Colonel Lum Wai-sing, a KMT officer.

As the students completed their training, some in China and some in Russia, they were assigned to air squadrons attached to the various KMT armies and were equipped with the heavier R-1.M-5 biplanes. Most of the flying students were from China's growing middle class, and many had studied abroad. The Russian advisers complained to Moscow that many of them had little or no knowledge or understanding of the national revolutionary movement and needed political education.

By June 9, 1926, when Chiang Kai-shek assumed full command of the land, naval, and air forces of the national revolutionary movement, the armed forces were already beginning to split into two groups, one virtually under the command of Borodin and General Vassily Galen, also a Russian, and the other loyal to Chiang. The pilots assigned to the Russian-controlled forces—particularly the IV Corps, under Hwang Chi-hsiang—became the first true Communist Chinese airmen.

In the fall of 1926, with about 200 aircraft between them, the IV and VII Corps of the KMT armies, the latter under Chiang Kai-shek's command, embarked on a northern campaign designed to defeat the war lords. The use of aircraft by the KMT forces in the field was a startling development, one that gave the national revolutionary movement enormous prestige as the armies advanced. Various Chinese war lords, particularly those in northern China, had previously used aircraft, and some even had small air forces. But the professionalism of the Russian-trained KMT pilots was feared and admired throughout China. The red, white, and blue twelve-pointed star insignia of the KMT Air Force, a device designed by Dr. Sun Yat-sen, became widely known. Often, the first warning of the coming KMT armies was a noisy low-flying pass by a flight of three R-1.M-5 biplanes on a combination reconnaissance and fear-building mission. The victories of the two KMT units were rapid, and the advancing forces took Wuchang in October, 1926; Hangchow in February, 1927; and Shanghai and Nanking in March, 1927. The split between the two army groups grew wider with each military success. Finally, in an attempt to "steal" the revolution, the Communists set up what they claimed to be the sole legal capital of the KMT Government at Hankow. Dedicated Communists from all over the world gathered there in the summer of 1927 to take part in what appeared to be the new government of China.

Reaction to the intra-army conflict on the part of Chiang Kai-shek's Nationalist forces had been immediate and violent. In the spring of 1927, at his new base in Nanking, which later became his capital, Chiang temporarily halted his northern march. For a month or more, it looked as if the two armies would fight each other. Each side flew reconnaissance missions over the other's territory with the aviation squadrons attached to their armies but avoided an air fight. Finally, late in June, the Nationalists presented an ultimatum to Hankow. Within weeks, the Hankow Government collapsed, with its members either joining the Nationalists or fleeing for their lives. On July 27, 1927, Borodin, Galen, and the other Russian advisers and instructors were expelled from China. It seemed a clear-cut victory for the Kuomintang, and Chiang Kai-shek believed he had eliminated the Communist threat to China.

But remnants of the Chinese Communist forces broke away

from the KMT Army and assembled at Nanchang under the leadership of Chu Teh, for more than the next thirty years military leader of the Chinese Communists. On August 1, 1927, Chu Teh was influential in creating the Red, or Communist, Army of China. Facing an uncertain future, he soon headed his forces south to set up a secure and protected base area. Other Communist forces headed for areas that would offer them some security, until there were at least eight Communist strongholds scattered over the country. The few aircraft remaining with the Communists were soon lost or destroyed, and their first semblance of an air arm disappeared.

As the Communists marched south, Chiang Kai-shek again headed his forces north and within a year took Peking, the old northern capital, renamed it Peiping ("Northern Peace"), and began to solidify the central government of China under the Kuomintang. In the meantime, Chu Teh and Mao Tse-tung, meeting for the first time, formed a partnership of control, settled their combined forces in Fukien and Kiangsi provinces, and later established their seat of government at Juikin. For the next six years, a constant state of civil war existed as the Communists, battered and all but beaten, were hunted down by the Nationalists in a series of extermination campaigns.

Cut off from the outside world and any military aid, the Communists survived by following their own classic methods of guerrilla warfare, primarily the avoidance of a military showdown with the Nationalists. They were able to pick up a few aircraft from local war lords, including Avro Avian IV trainers and some ancient French Breguets of World War I vintage, but none of those was suited for combat use. It was not until the fall of 1933 that any hope existed for a Communist air arm.

The central government's Nineteenth Route Army, under General Tsai Ting-kai, entrusted with a major part of the Fourth Extermination Campaign, moved its combat air units to western Fukien on August 5, 1933, for use against the Communists at Changting. However, riddled with Communist sympathizers, the Nineteenth soon revolted against the central government and offered to join forces with Mao Tse-tung. This would have more than doubled the Communist forces, giving them an air arm equipped with modern Curtiss Export Hawk II fighters, Loening-Keystone amphibians, and a complete flight-training school equipped with British Avro and Chinese-built Amoy trainers. Mao Tse-tung's failure to ally his forces with the Fukien Nineteenth Route Army is ascribed to instructions from Moscow. In any event, the Communists remained aloof as the Nineteenth formed its own "Federal Revolutionary Government of China" and made ready for war against the Nationalists.

Known as the Fukien Rebellion, the fighting lasted less than two months, during which time the Nationalist pilots virtually wiped out the Fukien air and ground forces in a series of air attacks. Fukien fighter aircraft defending the capital at Foochow were shot down in the air or destroyed on the ground, and some of the aircraft were captured by the Nationalists, thus eliminating the chance of having the Fukien planes fall into Communist hands.

It would be twelve years before the Communists would get another opportunity to obtain aircraft in numbers. But, in the spring of 1934, after the collapse of the Fukien revolt in January, the Communists faced the greatest trial of their history. Growing Nationalist power, and Chiang Kai-shek's need to rid the country of Communists once and for all, led to a massive military operation known as the Fifth Extermination Campaign. The plan was to ring the Communist areas with blockhouses and troops and then slowly close the ring. By the summer of 1934, the pressure was felt wherever the Communists turned, and Nationalist aircraft were continually on the attack. Finally, in desperation, the encircled Communists broke out in a series of assaults in October and November,

1934, and began a journey that lasted for more than a year to safer quarters in sparsely settled north-central China. Known to the Communists as the "Long March," it was later dubbed the "Long Capturing" by the Nationalists. The route led the Communist forces through some of the most difficult terrain in China. Of the estimated 300,000 Communist men, women, and children who started out, only about 40,000 completed the march. For a total of 368 days, 268 of which were spent on the trail, the marchers were continually pounded by Nationalist aircraft as they moved west and north. It wasn't until late October, 1935, that the major Communist group, under Chu Teh and Mao Tse-tung, settled in northern Shensi Province and later established a capital at Yenan. Their capital remained there until 1947.

It was during this period that Mao's forces obtained their first combat aircraft, a Nationalist Douglas O-2 MC4 general-purpose plane that had crashed behind Communist lines. It was named Marx, and although it was rebuilt and repaired, the three or four qualified pilots who remained with the Red Army could never get it off the ground.

The second Red Chinese aircraft was a former Nationalist Vought V-65-C1 that was flown over to the Communists at Yenan by its defecting pilot. More than seventy Voughts had been obtained by the Nationalists from the United States, and they were among the aircraft most widely used against the Communists along the route of the Long March. In honor of its Communist pilot, the Vought was renamed the Lenin and was used on communications and reconnaissance missions. It became a prime target for the Nationalists, and it wasn't long before Chiang Kai-shek's air force was avenged. On October 5, 1936, the Lenin was shot down by the Nationalist Air Force, and the Communist air arm once again was without aircraft. The Nationalists were proud of their victory, and to this day a photograph of the wreckage of the Lenin is on display at the Chinese military museum on Formosa.

The civil war and growing apprehension over Japan's plans for the future led to continual unrest in China. Chiang Kai-shek was kidnapped in December, 1936, by a group that wanted him to make peace with the Communists. Finally, the Japanese invasion of China in July, 1937, resulted in a semblance of cooperation—on the surface at least—between the Nationalists and the Communists. In August, 1937, Chu Teh and Chou En-lai, today Premier of Red China, flew to Nanking in a Nationalist transport for a conference with Chiang Kai-shek. The result was that both groups agreed to face their common enemy together. The Red forces were designated the Chinese Eighth Route Army, and the Kuomintang insignia, now a twelve-pointed star in a white field, was adopted for Red Army use and the few aircraft that were to be operated by the Communists. The Nationalists supplied the Communists with ammunition and money but no heavy equipment or aircraft as requested. A small number of obsolete Russian aircraft, clandestinely supplied by the Soviet Union and flown by volunteer Russian pilots, were made available to the Red Army. The Russians also trained a few Chinese pilots in their use. Polikarpov R-5 general-purpose biplanes were used for reconnaissance and field liaison, and a few bombers, reportedly Tupolev SB-2's, were also flown by Russian and Chinese crews in support of the Red Army. Captain Grigori Akimovich Kulishenko, killed when his bomber crashed after an attack on Japanese positions, became one of the first heroes of the Chinese Communist air arm.

The truce between the Nationalists and the Communists was an uneasy one, and while the Nationalists bore the brunt of the fighting against the Japanese for the next eight years, the Communists concentrated their attention on communizing the areas under their control, sporadically fighting the Japanese in northern China. During this period, the Communists repeatedly asked Chiang Kai-shek for aircraft to establish their own air force, but the idea was always rejected by the Nationalists at Chungking because of the ever-present threat of a renewed civil war. Clashes did occur, and on several occasions, Japanese forces stood by and watched as Chinese Communists and Nationalists fought each other, one action taking place along the Yangtze River in 1940. Flying the Russian aircraft remaining to them and a few captured Nationalist and Japanese aircraft through the World War II years, the Communists had no air force to speak of and little opportunity to use aircraft in action.

An airfield had been built at Yenan for use by visiting aircraft and the small number of transports available to the Communists. It was here, on July 22, 1944, that a small team of Americans arrived to analyze the military potential of the Chinese Communists. Once again, the Communists put in a strong bid for military aircraft, and this time the plea almost succeeded. Impressed with the Communists' battle claims and their plans for the future provided they received American fighters and bombers, the U.S. mission suggested that Chiang Kai-shek share some of his supplies with his old foes. The result was an immediate crisis between the United States and the Nationalist Government. The plan was dropped, and the Chinese Communists were cut off from any possibility of American aid.

But the stage was set for the greatest crisis of all, the end of World War II and the fighting between the two sides for control of the Chinese mainland. When the war finally ended, in September, 1945, China was in a state of chaos. The Communists had survived the fight with a larger area of control than ever before. The central government, however, was extremely war-weary, and the country suffered from inflation and political corruption. American policy, designed to unify and strengthen China, was based on cooperation between the two opposing forces, and repeated attempts were made to bring the two groups together in a coalition government.

To serve this end, American Ambassador General Patrick

J. Hurley went out of his way to encourage the Communists to negotiate with the Nationalists. His third request met with success, and Mao Tse-tung agreed to go to the Nationalist capital at Chungking to discuss postwar reconstruction with Chiang Kai-shek. Ambassador Hurley placed a former USAAF Douglas C-47 Dakota transport at Mao's disposal for use between Yenan and Chungking. On August 28, 1945, flying for the first time in his life, Mao Tse-tung, joined by Chou En-lai, the latter already an experienced air traveler, arrived in Chungking for the talks. The almost two months of discussion led nowhere, and on October 10 Mao left Chou in Chungking to represent the Communists and flew back to Yenan, taking his new aircraft with him. Bedecked with Chinese Communist markings over its original American drab finish, it became Mao's personal transport and mobile headquarters during the ensuing civil war.

While the conversations continued at Chungking and Yenan, the race was on to see which side could occupy and control the greatest amount of territory. In a matter of weeks, the Chinese Communists fanned out from northern Shensi, illegally occupying Japanese-held territories and accepting surrenders from the Japanese and forces of the puppet government of occupied China. At times, the Communists posed as Nationalists to take over the equipment and supplies. They were soon in close proximity with the Russians and Outer Mongolians who, only a few weeks before, had invaded the Japanese puppet state of Manchukuo, now once again Chinese Manchuria.

It wasn't until December, 1965, with the publication in Moscow of a World War II history, that the Russians finally admitted that they had gone against the terms of the Japanese surrender to aid the Chinese Communists in Manchuria, but it was obvious to the Nationalists as soon as it happened. The Chinese Communist armies had never before made direct contact with the Russians, and Chinese forces were uneasy as they entered Manchuria. The outcome was beyond their wildest dreams. While the Soviets prevented the landing of Nationalist troops along the coast, all the weapons, equipment, supplies, and aircraft of the Japanese Army in Manchuria were turned over to the Chinese Communists. For the first time, they possessed modern arms in quantity, but they did not know how to use them. Japanese and former Manchukuoan pilots and technicians, not knowing what to do next, signed up as mercenaries with the Chinese Communists by the hundreds and both manned the aircraft and trained the Chinese in their use.

The new Red Chinese Air Force was theoretically the strongest air arm in northern China. Modern in every way, it was made up of Nakajima Ki.43 Type 1 Hayabusa and Ki.84 Type 4 Hayate fighters of the 48th, 104th, and 204th Japanese Army fighter regiments formerly stationed in Manchukuo; Nakajima Ki.44 Type 2 Shoki and Kawasaki Ki.61 Type 3 Hien fighters stationed in northern China; and more than 100 Mitsubishi Ki.51 Type 99 ground-attack and Kawasaki Ki.48 Type 99 two-engine light bombers captured in Manchuria. Of more immediate importance to the Communists were the many transports and trainers obtained at Mukden and Harbin, the major Manchukuoan Japanese Army Air Force bases. Mitsubishi Ki.57 Type 100, Tachikawa Ki.54c Type 1, and Nakajima Ki.34 Type 97 transports flown by Japanese pilots entered into Red Army service immediately, providing the field armies with the first direct air communication they had ever enjoyed. Tachikawa Ki.55 Type 99 and Manshu Ki.79 Type 2 advanced trainers, as well as Tachikawa Ki.54a Type 1 two-engine advanced trainers, were taken over in droves by the Chinese at Mukden, and by October, 1945, they provided the Communists with all the equipment they needed for modern flight-training schools set up in Manchuria. Many of these former Japanese trainers remained in PLAAF service for many years, with some of them in service until the early 1960's. An example of the Tachikawa Ki.55, long the standard advanced trainer of the Red Chinese Air Force, is still on display in the military museum in Peking.

Apparently, the new air force was first used tactically in May, 1946, when three unmarked fighters of Japanese origin came in low and strafed a Nationalist armored train north of Mukden, in Manchuria, killing at least 100 men. There is little question that the pilots were Japanese, for the Chinese did not yet have pilots of their own capable of handling such high-performance aircraft. When the raid took place, the military situation in northern China and Manchuria was in flux. The Nationalists had begun to move in and reclaim the territory first occupied by the Communists. A cease-fire effected on January 13 did not hold up, and on April 15 the Communists went on the offensive again after refusing to take part in a coalition government. Nationalist airlifts and military trains, backed by superior air power and well-trained troops, soon began to put the Communists on the defensive, and it seemed that Chiang Kai-shek's forces would win the day. But once again, a cease-fire was negotiated–agreed to by the Nationalists because they risked losing all further American aid if they continued fighting–and on June 6, 1946, the Communists were saved from almost certain defeat.

The Chinese civil war formally began in July, 1946, when the Communists announced the creation of the People's Liberation Army (PLA) and the Nationalists actively began to prepare for the open hostilities they knew were inevitable. The Nationalist Air Force was moved into position in northern China. The truce was almost exploded when the Nationalists demanded the return of one of their Consolidated B-24J Liberator bombers originally supplied by the United States and currently in Communist hands. The aircraft had landed at the Red capital at Yenan in June, 1946, providing the Communists with their first four-engine bomber. The Communists claimed that the pilot had defected, and the Nationalists promptly countered with the statement that the aircraft had landed at the Communist base because it had run out of gas on its flight from Chengtu. The Nationalist Air Force followed up the demand with a raid by six planes in August, which the Communists claimed strafed the city in addition to completely destroying the B-24. The Communists asked for punishment of the men involved and reparations for the damages. The question was never resolved as other incidents piled up and full-scale fighting came closer to reality.

In November, the Nationalists appeared to have the upper hand in the Battle of Kalgan, and early in 1947, Chiang's armies were getting ready to assault the Communist capital itself. As the Nationalist forces pressed closer to Yenan, the Nanking Government confidently announced the defeat of the Communist armies. Plans were made for a strategic Communist retreat. John Roderick, Associated Press photographer in Yenan at the time the Nationalists began to close in, asked Mao, "Have you given up?" Mao's reply was, "Mr. Roderick, I invite you to visit us in Peking two years from now." As the Battle of Yenan unfolded, the Communists pulled back; by March of 1947, Nationalist losses were mounting so rapidly that the capture of Yenan lost its significance. By May, the initiative slowly but surely began to pass to the Communists, never to be regained by the Nationalists.

In December, 1947, the Communists went on a winter offensive in Manchuria supported by an air force of about 200 operational aircraft. One by one, the Nationalist garrisons fell. As the momentum grew, the Communists began to win stunning victories. Yenan was recaptured the following March, and by November 23, 1948, a huge Red pincer movement caught the Nationalist northern armies in a trap. By the end of January, 1949, all of China north of the Yangtze River was in Communist hands.

In preparation for their planned crossing of the Yangtze, the Communists established a flight-training school at Harbin,

Manchuria, where Japanese and Russian instructors trained the Red Chinese in the use of about 100 Japanese-built light bombers. Gasoline was in short supply, and Nationalist Air Force bases and gasoline stocks became prime targets for the PLA advance. Captured Nationalist pilots and aircraft were quickly incorporated into the growing Communist air arm.

At the beginning of the Red advance, the Communists rarely used their aircraft in combat. In fact, the Nationalists, regarded the PLA as having no air force at all. But the Red air arm was becoming more formalized, if only in a loose manner. Early in the fighting, Red aircraft carried no markings of any kind; removal of the Japanese insignia was the only official requirement. But by the end of 1946, a new insignia—a red star with the character for China at its center—began to appear. It was hastily painted over the Nationalist blue-and-white star insignia as more and more government aircraft fell into Communist hands. Between July 1, 1946, and January 31, 1949, the Communist Chinese claimed that they had captured 86 serviceable Nationalist aircraft, and that 15 of them had been flown over to their side by defecting Nationalist pilots. This continuing transfusion of trained men and modern American aircraft both strengthened the Communist forces and reduced the effectiveness of the retreating Nationalist air forces.

On the night of April 20, 1949, the PLA crossed the Yangtze River, and China was all but lost to Communism. On April 24, the Red forces took Nanking, and Shanghai fell on May 25. The Nationalists, anticipating the Communist victory, fled to Formosa, taking as much of the remaining Chinese Air Force with them as possible, and established the capital of the exiled Republic of China at Taipei. By October, 1949, the civil war was virtually over, and Peking was declared the capital of the People's Republic of China. By December, the PLA had reached the border of French Indo-China and by the following April had invaded and occupied Hainan Island.

In the last twelve months of fighting, the Communists had captured hundreds of Nationalist aircraft, bringing their total number of planes to more than 500. Only about half of them were airworthy, and by the end of the fighting, the PLA air arm had about 150 serviceable fighters and between 75 and 100 transports and bombers, most of them flown by ex-Nationalist pilots. In addition, over 1,400 trained Nationalist aircraft technicians were captured at Shanghai, at last giving the rapidly growing Communist air arm the ability to service properly its growing inventory of aircraft. The new aircraft were largely of American origin, and the Red air arm was faced with the multiple problems of maintaining and flying a wide variety of aircraft. The PLAAF was formally organized late in 1949 under the command of General Liu Ya-lou, who had been chief of staff of the Fourth Field Army. An attempt was made to standardize equipment, with North American P-51D Mustangs becoming the basic PLAAF fighter, backed up by Republic P-47D Thunderbolts and De Havilland DH-98 Mosquitoes, the latter originally purchased by the Nationalists in Canada. North American B-25H Mitchells formed the backbone of the new Communist bomber units. Scores of Douglas C-47 Dakota and Curtiss C-46 Commando transports also entered Red service, and many of them are still in use on the Chinese mainland. Boeing PT-17 Kaydet trainers were picked up and sent to the PLAAF flying schools, as well as some helicopters that the United States had provided the Nationalists.

At this point, the entire Red Chinese military needed drastic reorganization. General Liu reported that the PLAAF strength early in 1950 was only about 100 usable aircraft, including civil transports, and that most of the captured Nationalist aircraft were in poor condition. Once again, the Russians, who had maintained a hands-off policy toward the civil war, came to the rescue, but now they were virtually given carte blanche. The Russians had sent a military mission to Peking in July, 1949, when they recognized the possibility of a Communist victory on the mainland. The Soviet Air Force was represented by a Colonel Voroshilov, sent to China to supervise a pilot training program. Preliminary work was started on a primary flight program, but Voroshilov subsequently reported to Moscow that the training of Chinese pilots would be too costly and time-consuming and that the need did not exist, as the Chinese were assured of victory. Within a few short months, this entire concept changed. On December 16, 1949, Mao Tse-tung went to Moscow as a Communist hero and ultimately signed an agreement for Russian aid at all levels, military and civilian, to bring Red China up to the status of a world power. By April, 1950, Russian technicians and advisers were swarming into China. One of their prime goals was to discard the haphazard structure of the air arm and create a modern force along Russian lines, including regiments, divisions, and air armies. Russian equipment was rushed into China, and PLAAF fighter units were revitalized. The equipment for the fighter regiments was standardized, with units flying about 200 Lavochkin LA-9, Lavochkin LA-11, Yakovlev YAK-9P, and North American P-51D fighters. Jets, in the form of a few Mikoyan MiG-9 fighters, were also introduced to the PLAAF.

Bomber regiments were formed and flew second-line Petlyakov PE-2 light bombers and Tupolev TU-2 attack bombers as well as the North American B-25H Mitchells already in use. The TU-2 bombers took part in a flyby over Peking in the summer of 1950 in commemoration of the founding of the Chinese People's Republic (October 1, 1949). The Russians also provided more than 100 Ilyushin IL-2m3 and IL-10 ground-attack bombers, as well as Lisunov LI-2 and Ilyushin IL-12 transports, which joined the ex-Nationalist C-46 and C-47 transports being used by the Red Chinese under Russian tutelage. The entire structure of the PLAAF was upgraded, and an air-force academy was established at Sian in Shensi

Province. Single-engine Yakovlev YAK-11 advanced trainers and twin-engine YAK-16 trainers joined the former Japanese Ki.54a and Ki.55 trainers still in service. Yakovlev YAK-12 light planes were supplied for liaison and communications duties, along with Polikarpov PO-2 utility biplanes and MI-1 helicopters. It was also during this period that the standard red star-and-bar insignia of the PLAAF made its appearance, remaining in use until the present day.

The Russians had hardly started work on their reorganization of the PLAAF when the Korean War broke out (June 25, 1950). After an initial series of setbacks, U.S. and Republic of Korea (ROK) forces repulsed the North Korean invaders and wiped out the North Korean Air Force. By late October, American troops were racing headlong for the Yalu River border of Red China. Chinese intervention, an unknown factor, was considered a dim possibility, although it was known that the Communists had embarked on a build-up of forces in Manchuria. In fact, at a conference with the Russians and North Koreans in August, the Chinese Communists had agreed to cross into Korea provided the Soviets supplied Red China with modern arms and aircraft. Within a matter of weeks, the Russians, under Stalin's direct orders, began deliveries of Mikoyan MiG-15 jet fighters to China and started training Chinese pilots in Russia and at Mukden, renamed Shenyang by the Communists. The jet force expanded rapidly, backed up by a conventional fighter force of about 160 Lavochkin LA-11 and Yakovlev YAK-9P aircraft stationed in Manchuria, facing the North Korean border. The first confrontation took place on November 1, 1950, when six Chinese MiG-15 jets jumped a flight of four USAF F-51 Mustang fighters over Namsidong in North Korea. This unequal combat was inconclusive, and the Mustangs escaped. A week later, on November 8, the first aerial combat between jets took place when four American Lockheed F-80C Shooting Stars met four MiG-15's in the air and brought down one.

The next day, an American RB-29 was lost to MiG-15 fire, and the air war escalated rapidly.

By November 26, 1950, when the ground forces of the so-called Chinese People's Volunteer Army crossed the Yalu into North Korea, air combat between Chinese and American aircraft was becoming a daily event. In one short year after the end of the Chinese civil war, Red China was at war in fact, if not in name, with the United States. The MiG threat became so acute that the USAF quickly dispatched its newest service fighters, North American F-86 Sabres, to Korea, the first models engaging MiG's in combat on December 17, 1950.

The surprise entry of Red China into Korea caught the American and ROK forces off guard, and by January 4, 1951, the Communists had recaptured most of North Korea and had entered the South Korean capital of Seoul. The fighting was intense, and by June, 1951, the front had become stabilized along the original border of the 38th parallel. In spite of the Chinese attempts to break the deadlock, both on the ground and in the air, American power held the Chinese in check while truce negotiations took place for the next two years.

As the Chinese armies raced south late in 1950 and early in 1951, the "volunteer" air force followed, attempting to set up air bases throughout North Korea, placing Chinese bomber and fighter forces within range of major American bases in South Korea. U.S. Seventh Fleet units off Korea were buzzed by Ilyushin IL-10 attack planes of the PLAAF, and on one occasion an ex-Japanese float plane of the Chinese People's Liberation Army Naval Air Force (PLANAF) attacked and was knocked down instantly. It was soon apparent that American supply lines and the sole South Korean port facility at Pusan were within Chinese bombing range. Escorted by MiG-15 fighters, the Red Chinese force of about seventy-five Tupolev TU-2 attack bombers committed to Korea could easily raid these targets.

The American plan to bring the war directly to China, including the use of atomic bombs, if South Korea was attacked by the Red Chinese Air Force was clearly stated to the Indian ambassador in America, and the warning was quickly transmitted to Communist China. The Red Chinese, in turn, let it be known through diplomatic channels that American bombing of Manchuria or the use of American aircraft over China proper would result in Chinese bombing attacks on South Korea as well as on U.S. bases in Japan.

The result was an unwritten binding but limitation on the opposing sides that was carefully observed throughout the remainder of the war. Although Chinese bombers never crossed the 38th parallel to raid the American concentrations of men and materials, the introduction of twin-jet Ilyushin IL-28 medium bombers at the PLAAF's Manchurian bases in November and December, 1952, made the Communist Chinese threats all the more meaningful. Moreover, Chinese airfields set up in North Korea became prime targets for American air power in order to remove the threat of such raids. Only when the ground facilities of the Chinese Air Force had been virtually wiped out south of the Yalu and the Chinese air arm had been completely forced back into its Manchurian bases early in 1953 did the U.S. pledge of "hot pursuit" of Chinese aircraft across the Chinese border, as well as the threat of an atomic attack on Chinese cities, have true meaning. When the Red Chinese faced the fact they were no longer able to retaliate, truce agreements were promptly reached, and the Korean War came to an end in July, 1953.

The U.S.–Chinese understanding was not announced to the American public at the time, nor was it revealed by the Communists. Its restrictions prevented both sides from making full use of their air forces, and the air war over North Korea took on an unrealistic character, providing the Red Chinese Air Force with a sanctuary in Manchuria and the American ground forces and port facilities a haven south of the 38th parallel. The Red Chinese air arm had full use of a massive complex of air bases just across the Yalu River border for the full thirty-two months of the aerial war, between November, 1950, and July, 1953. Safe behind an invisible wall past which American pilots could not fly, the Communists built up their air force and swarmed into the North Korean skies. Their mission was uncomplicated, as their only task at this stage was to defend the power plants, bridges, and supply lines along the Yalu River, through which thousands of reinforcements and massive shipments of materials flowed. The American job was infinitely more difficult. Flying at extended range, often low on fuel, and with limited combat time available, American fighters had to clear the air for American bomber attacks on Red Chinese facilities and airfields along the southern edge of the Yalu and in North Korea.

The odds favored the Communists. Red Chinese pilots could calmly mount their MiG-15 fighter aircraft at any of the four major Manchurian air bases—at Antung, Tatungkow, Takishan, or Tapao; climb to altitude over Manchuria without fear of interference; cross the river at Suiho or Sinuiju at about 48,000 feet; and come out of the sun at the American F-86 Sabre fighters just reaching "MiG Alley" after a long flight. The fighter passes would be made at the formerly unheard of closing speeds of more than 1,200 mph, and the combat was bitter. Red Chinese formations of as many as 200 MiG-15 fighters would engage much smaller American forces. After the initial pass, the opposing fighters would war in individual dogfights.

The first of the major air battles began in March and April of 1951, soon after Russian deliveries of MiG-15 fighters were made in quantity. By May and June, formations of as many as 180 Chinese MiG-15 fighters were encountered, flown by action-ready Russian-trained pilots. PLAAF strength grew rapidly, and by July, 1951, American intelligence estimated that the Chinese had 1,050 aircraft on hand in Manchuria for use over Korea. Of these aircraft, 595 were fighters—445 of them jets—while the remaining 150 were older Russian and former Nationalist piston fighters. The remaining strength included 175 bombers and ground attack aircraft, 100 transports, and 180 reconnaissance and training planes. Photo-reconnaissance of the Manchurian air bases revealed as many as 400 fighters at a single field. The Red Chinese pilots were as yet untrained in combat, and their losses mounted rapidly. Yakovlev YAK-17UTI jet trainers were supplied to the Red Chinese training schools, and a few Lavochkin LA-15 jet fighters were also sent to China, although they do not appear to have been used in combat over Korea. As the year wore on, the Chinese pilots gained in proficiency; by the summer of 1952, the American kill rate began to drop appreciably. Flying advanced MiG-15bis fighters supplied by the Russians in 1952 and 1953, the better Chinese pilots became bolder, and rather than avoid combat as they had often done in the past, they aggressively pursued F-86 Sabres that were heading back to base as their fuel ran low.

In spite of their advantages and the seemingly endless supply of Russian-built MiG-15 fighters, the Chinese failed in their mission and never achieved air superiority in North Korea. The MiG's were credited with seventy-eight confirmed Sabre kills and twenty-six probables, as well as downing forty-two USAF F-80, F-84, and F-51 fighters. Only sixteen B-29 and six B-26 bombers were lost to the MiG's, and American bombers continued to raid the North Korean installations. The Red Chinese losses, on the other hand, were staggering. While the USAF officially claimed that 841 Communist MiG-15 fighters were shot down in combat, the actual count was well over 850. Mao Tse-tung's own son, Mao Anying, an air-division commander of the PLAAF, was one of the first Red Chinese pilots to be downed by American fighters over Korea, in 1950. His death was a great personal loss to Mao. It was only the beginning. Other Chinese losses—such as crash landings in Manchuria, head-on collisions observed over the airfields as the frantic and frightened Chinese rushed back to their bases, and MiG's that blew up in the air after suffering unconfirmed hits—added an estimated 400 losses to the toll. Without question, hundreds of MiG's were lost in training, for many unskilled Red pilots were forced into the jet age by Mao Tse-tung's decision to invade Korea. As jet-training time was reduced by the Chinese to meet the requirements for more pilots, the degree of skill dropped further. While the fighting continued in the spring and summer of 1953 and the seemingly endless truce talks continued at Panmunjom, the U.N. and American MiG kill rate climbed. Although the Chinese Communists never revealed their losses, instead concentrating on their successes and lauding their Korean "aces," such as Li Han, Chang Chi-wei, and Wang Hai, the combat reports and gun-camera records of the American Sabre pilots, as well as the rapid drop of PLAAF activity over North Korea in July, were revealing.

By June, it had become more than obvious that the Red Chinese air arm was in deep trouble. During that month, American pilots confirmed the downing of seventy-seven MiG fighters and reported eleven probables, with no loss of U.S. aircraft in air-to-air combat. An additional forty-one MiG's were reported damaged, and the actual Chinese loss probably totaled more than a hundred aircraft. The message could hardly have been lost on the PLAAF regimental-staff personnel in Manchuria as they awaited the return of their aircraft, and it certainly was clear to the ever-changing cadre of PLAAF pilots being fed into the skies over North Korea to engage the USAF.

In July, the carnage went on, but in a matter of days, Red Chinese aerial activity dropped off rapidly as the truce talks came closer to conclusion. On July 20, the PLAAF scored its last Sabre kills of the war when two of the American fighters were downed by six MiG-15 fighters obviously flown by skilled Chinese pilots. But by July 22, the day of the last

American MiG kill of the war, the reported American score for the month was thirty-one MiG's, with the USAF air-to-air losses standing at two. The total Red Chinese loss during the Korean fighting exceeded 2,000 fighters, and the cost must have given the Russians second thoughts about their involvement on more than one occasion. A nationwide "arms donation" campaign conducted in the People's Republic of China to collect money for the purchase of aircraft and arms, running from June, 1951, through May, 1952, had provided well over $200 million. Although it contributed to the enthusiasm of the civil population for air power and the PLAAF, with the bulk of the contributions coming from the cities, even this effort wasn't enough to be of any significant help. The PLAAF aircraft losses put Red China deeply in debt to Soviet Russia.

MiG-15 fighters were not the only Red Chinese aircraft used over Korea, and by the fall of 1952, about 2,100 Communist planes had been involved in the fighting, of which about one-third were a variety of other aircraft types. Tupolev TU-2 attack bombers—escorted by Lavochkin LA-9 and LA-11, Yakovlev YAK-9P, and, occasionally, MiG-15 fighters—were used over North Korea on tactical missions late in 1951 and early in 1952 against both mainland targets and the Korean offshore islands. The Russians continued to supply the PLAAF with a mixed bag of aircraft types, and right up to the end of the fighting the Soviets were still delivering obsolete LA-9 and TU-2 piston-engined fighters and bombers. In February, 1953, American estimates placed the Chinese inventory of combat aircraft at 1,400, including 250 piston fighters and 220 piston bombers, a substantial increase over past years. The remaining combat force was made up of 830 jet fighters and an estimated 100 jet bombers of the IL-28 type. Ilyushin IL-10 aircraft were used on ground-support missions, and by the spring of 1953, PLAAF transport companies reached North Korea flying former Nationalist C-46's and C-47's as well as Russian-built Lisunov LI-2's and Ilyushin IL-12's. One transport company, with women pilots flying C-47's, was widely publicized in the Red Chinese press. About 150 of these second-line combat and transport aircraft were confirmed lost by the Red Chinese to American fighters, and the actual count was probably much higher. By the time the last aerial engagement of the Korean War took place—on the afternoon of the day of the truce, when a PLAAF IL-12 transport on a courier flight was caught by an American Sabre pilot over North Korea just before it headed back across the Yalu River to Manchuria—the PLAAF was thought to have a service strength of 2,868 aircraft. In the last year of the Korean fighting, the PLAAF had more than doubled its maintenance and ground-support facilities and had added eight new flight-training schools, bringing the total to eighteen, with the major PLAAF air academy at Sian. However, these advances were misleading, for the Korean experience had been a costly one.

The closing weeks of the war led to a growing American suspicion that the PLAAF had suffered an even greater defeat than was first imagined. If negotiation was the historic Chinese way of gaining time until conditions improved, then the Red Chinese willingness to reach a truce in June and July and their obvious concern over the delays of the South Korean Government could mean that more than a desire to establish peace in Korea lay behind their change of heart. American speculation over the possibility of a harsh settlement—based on the assumption that the Chinese Communists were no longer able to contain the U.N. forces in Korea or to defend mainland China against invasion or aerial assault—came to an abrupt halt with the signing of the truce on July 27, 1953. The final agreement possibly saved the People's Republic of China from military ruin and eventual defeat. Red China's air forces .had, in fact, been virtually neutralized, and most of mainland China would have been helpless in the face of an American air attack. With peace in Korea, the moment of justifiable military action for the United States quickly passed, and world attention focused on other parts of Asia.

The reality of the truce created a new series of problems for the air forces of Red China, for now the lines were clearly drawn. The People's Republic of China and the military power of the United States stared at each other across the 38th parallel and along the Red Chinese coastline. In the background lay Taiwan, an eternal thorn in Red China's side. No longer could Communist China play the role of a good neighbor providing a "volunteer" military force for a friend in need; for that matter, neither could the United States. While many nations of the U.N. had participated in the Korean "police action," embroilment in a fight between opposing Chinese governments was another matter. The end of the fighting in Korea relit the ever-sputtering fuse of the Chinese civil war, with U.S. forces standing alone in the middle, both figuratively and physically. The U.S. Seventh Fleet, posted to Chinese waters on a watchdog mission when the Korean War began, coursed along the coast of Communist China. Its announced purpose was to prevent a highly touted Nationalist invasion of the mainland from Taiwan, but in reality its duty was to discourage any Communist Chinese attempt at the reverse objective. The days immediately following the Korean War were nervous ones for the fleet, since its ships were vulnerable to a possible Red Chinese air attack. A series of inconclusive aerial engagements over the Yellow Sea between USAF Sabres spoiling for a fight and equally aggressive PLAAF fighters, resulting in the loss of at least one MiG-15, only added fuel to the burning question of what Red China would do next. Knowledge of the operational strength of the Chinese air forces, particularly the types and numbers of aircraft on the airfields along the coastal perimeter, was of critical importance to American plans for the future. The answer would be found in a matter of weeks as a result of a flamboyant display of American brinksmanship and a demonstration of power that left the Communist Chinese determined to rebuild their strength in the air and close their aerial borders to intruders for all time.

The question of Communist China's capability to attack the American surface fleet was important enough for Washington to take a calculated risk in order to test Chinese intentions. The Korean truce was barely a month old when a special American naval task force, built around a core of three carriers, headed for the East China Sea to assemble for a daring overflight of mainland China. The purpose of the mission was to find out what air power the People's Republic had stationed within range of the China seas and the straits of Taiwan, and the method selected was photoreconnaissance of airfield installations.

When the force went on station, the entire fleet was put on general quarters for two days. Pilots were informed that interception and retaliation by Red Chinese aircraft were anticipated and expected. When the carrier planes were launched, the fleet braced itself for a possible Chinese aerial attack. An estimated 300 American naval fighters and photoreconnaissance aircraft took part in the overflight, and as the massive formations raced along the coastal areas of the mainland provinces of Chekiang and Fukien, they found themselves in open skies. No PLAAF aircraft came up to meet them, and visual checks of the airfields below indicated that only a few planes were scattered around the fields. Evaluation of the reconnaissance photographs confirmed the fact that the Red Chinese air forces had almost no aircraft assigned to the area. The most vulnerable border of the People's Republic of China was undefended, and there was virtually no physical evidence of the modern jet-age air force that Mao Tse-tung had so eagerly thrown into the skies over North Korea. The dramatic show of strength had provided an answer. The air forces of Red China posed no threat to American plans. The Seventh Fleet remained in position. Communist China could not match American air power on her own borders and probably would not make any aggressive moves for some time.

Communist China seemed to be contained, and only a herculean effort, coupled with newly expanded programs of Soviet aid, could alter the balance. To add to the problems of the Red Chinese, their chief Soviet supporter, Joseph Stalin, was now dead, and the Sino-Soviet relationship would need rebuilding. But once again, the Russians came to the aid of their ally and provided the help the PLAAF needed to embark on a re-equipment program to bring it in line with the requirements of the future.

CHAPTER II

THE TIGER GROWS STRONG 1954–64

When Nationalist Captain Chen Wai-sheng and his wingman banked low over Shanghai in their RF-84F Thunderflashes to photograph Red Chinese installations early in 1956, they suddenly found themselves in a sky full of MiG's. Nearly forty People's Liberation Army Air Force fighters attacked them at once, and the two Taiwan-based pilots faced almost certain death in their unarmed reconnaissance planes. With a bluff their only alternative, they bore down on their foe and scrambled low over the city. Almost at once, Captain Chen found himself closing in on one of the PLAAF MiG's, and at 600 yards he fired the only thing at his command—his camera—and got away. When the film was developed back on the island of Taiwan, the fighter that revealed itself looked different from the MiG-15's met by the USAF over Korea and by the Nationalist Air Force over the Quemoy straits. The aircraft in the photograph was a MiG-17, and the camera record was the first evidence that this advanced version of the MiG-15 was in PLAAF service. It was but another demonstration of the massive Soviet aid poured into Red China after the Korean War, which built up the Chinese Communist air forces to the point where they felt they could gain a decisive victory over the annoying Nationalist redoubt on Taiwan.

The first indication of the new Red Chinese build-up in the air came in the summer of 1953, when U.N. prisoners were routed through Pyongyang, the capital of North Korea, on their way to Panmunjom in "Operation Big Switch," the post-war exchange of Korean War prisoners. Red Chinese Tupolev TU-2 bombers were spotted illegally landing at airfields in the area (in violation of the recently signed Korean truce agreements), along with a number of Russian-built twin-jet bombers of advanced design, later identified as Ilyushin IL-28 medium bombers, first spotted across the Yalu in December, 1952. Within a few months after the Korean War, the Russians had once again started to reorganize and revitalize the PLAAF and, in so doing, had created the largest tactical air force in the Orient.

The PLAAF was organized along Russian lines. The basic unit was the air regiment, generally made up of three air companies, with each company consisting of three or four aircraft squads. Three, sometimes two, air regiments made up an air division, a large unit that was given a specific duty and geographic assignment. By the end of 1953, the PLAAF consisted of eighty air regiments, assigned to twenty-six air divisions; three special air regiments; five paratroop divisions; and a naval air arm, designated as the 6th Air Division. Deliveries of new aircraft increased rapidly; by late 1955, the Chinese Communists had almost 4,000 aircraft, of which at least 2,000 were fighters and the remainder bombers, transports, and support aircraft. The PLAAF air academy at Sian, along with the many other training schools, produced a substantial number of new pilots, and the PLAAF became widely recognized as the elite corps of the People's Liberation Army. A new uniform, incorporating the Russian flair for color and decoration, also made its appearance, and the PLAAF soon found itself in a position where it could pick and choose its candidates for pilot training from the cream of China's youth.

The forced expansion of the Red Chinese Air Force placed ever-growing demands on the training facilities in China, particularly for multiengine and twin-jet bomber aircraft. To ease the strain, pilot and crew training was initiated at Tashkent, Kiev, Novosibirsk, and Vladivostok in Russia in 1955, and Chinese training establishments were staffed with Russian instructors and advisers as well as with new MiG-15UTI and Ilyushin IL-28U trainers.

The new equipment supplied by Russia was a step ahead of that used by the PLAAF in Korea. Initial concentration was on fighters to defend the mainland against possible Nationalist or U.S. attempts to bomb the rapidly expanding industrial complexes being developed by the Chinese Communists. While Korean-vintage MiG-15 fighters remained in service, along with some piston-engine Lavochkin LA-9 and LA-11 fighters for areas where the Red Chinese would not directly face the Nationalists, the new deliveries were made up of MiG-15bis fighters, followed in late 1954 by the advanced MiG-17, a more powerful and faster version of the basic MiG design. By 1955, more than 500 of the new fighters were in PLAAF service, and the Red Chinese were set once again to challenge the Nationalists for control of Taiwan.

While the fighter forces were being strengthened, the Chinese concentrated on developing the PLAAF bombing capability. The first order of business was to create tactical- and strategic-bombing forces, neither of which had been effectively used by the Red Chinese in Korea. The tactical force was built around the Ilyushin IL-28 medium bomber, and additional deliveries were made in 1953. By 1956, more than 250 of these jet bombers were in use, and at the end of the 1950's, there were about 700 IL-28 and older TU-2 bombers in service.

The most significant addition to the PLAAF in this period was the creation of a strategic-bombing force that brought all of Taiwan, as well as the American bases in Japan, Okinawa, the Philippines, and Korea, within easy range of the Red Chinese bombers. By a strange twist of fate, the Soviets supplied the Chinese with the same type of four-engine aircraft used by the USAAF strategic-bombing forces to bring Japan to her knees in World War II and to bomb the dams along the Yalu River during the Korean War. This was the Tupolev TU-4, a Russian imitation of the Boeing B-29 Superfortress, re-created by Tupolev from American examples that force landed in Russia during World War II and copied right down to the fire-control equipment and the powerful radial engines. The first Red Chinese TU-4 was delivered in late 1951, but it wasn't until 1954, when a long-range bombing school for Red Chinese pilots and crews was in operation at Kazan in the Soviet Union, that the force was a reality. By the summer of 1955, a PLAAF long-range bomber air division, consisting of two air regiments with almost a hundred TU-4 bombers, was established at Peking, and the Red Chinese had the strategic-bombing capability to strike at Taiwan.

Other Russian aid consisted of an assortment of Antonov AN-2, Lisunov LI-2, and Ilyushin IL-12 transports, as well as Yakovlev YAK-18 primary trainers for the Chinese schools and Mil MI-4 helicopters for the PLA's tactical-assault training program.

The PLANAF, the Red Chinese Naval Air Force, was established during this period for coastal-patrol and -defense duties, and in 1955 naval pilots were trained at the Fifth Aviation School at Tsinan, Shantung Province. Consisting of only one air division, the naval force was assigned MiG-15bis fighters for coastal-defense duties, as well as a number of Tupolev TU-2 attack bombers and Beriev BE-6 flying boats for patrol duties. Eventually, the navy received MiG-17 fighters and a few Tupolev TU-4 strategic bombers, modified for long-range reconnaissance service, as well as a reported fifty to sixty Tupolev TU-14 twin-jet land-based patrol bombers in the late 1950's to replace the obsolete TU-2.

From the end of the Korean War to the present day, the Red Chinese air forces have never enjoyed a respite from combat duty. The failure to take the offshore islands of Quemoy and Matsu in 1949 or 1950, not to mention Taiwan, remained a thorn in the side of the Communist hierarchy. The result was that the Chinese civil war never really came to an end and is still a fact of Chinese life. In the early 1950's, Nationalist bombers often raided the mainland to forestall Communist invasion build-ups, and Red interceptors were posted along the coastline to provide air cover for the assembled invasion fleets. The consequence was a tinderbox border stretching for thousands of miles, constantly patrolled by Communist fighter planes. This explosive condition resulted in a number of incidents. Nationalist and American reconnaissance planes, the latter patrolling the coast of China as part of the U.S. commitment to protect the Nationalist position on Taiwan, often made sight contact with the Red fighters. In July, 1954, air combat once again took place between U.S. and Red Chinese fighters, although both sides used aircraft of World War II vintage. Touchy PLAAF pilots, flying obsolete Lavochkin LA-9 fighters near the island of Hainan, off the south coast of China 600 miles away from Taiwan, shot down a British Cathay Pacific airliner that had inadvertently crossed into Red Chinese airspace. The Communists claimed that they thought it was a Nationalist transport. On July 25, three U.S. Navy F4U Vought Corsair fighters from the carrier *Philippine Sea*, searching for survivors, were jumped by a flight of LA-9 piston fighters; in the resulting dogfight, two of the Chinese fighters were shot down.

By the beginning of 1955, the PLAAF and the Nationalist air forces, the latter brought up to modern jet standards by American aid, were facing each other across the Quemoy straits, each side eager to demonstrate its ability to best the other in the air. Jet flying time across the open water was only fifteen minutes, yet the two forces were worlds apart. In the spring of 1955, the Red Chinese had a network of some forty air bases within range of Taiwan and the offshore islands, and preparations were being made for the final push to conclude the Chinese civil war with a Communist victory. Early in January, flights of fifty to seventy-five MiG fighters and IL-28 bombers sporadically attacked the Tachen islands (off the mainland some 200 miles north of Taiwan), which the Nationalists finally evacuated under fire in February. But the threats of Nationalist retaliation by Republic F-84G Thunderjet and North American F-86 Sabre fighters and the high PLAAF crash rate during the raids brought these scattered attacks to a close along the straits. The Nationalists retained air control over the narrow belt of water for the time being. Aerial clashes over the straits were infrequent but damaging to Communist morale. On October 15, 1955, a single F-86F of the Nationalist Air Force fought off twelve Red MiG's and brought one down in the process. The next summer, on July 22, 1956, the Nationalists shot down four MiG-17 fighters and damaged two more. Occasional clashes took place throughout 1957, but it wasn't until the summer of 1958, after the Communists had had a chance to build up their combined forces for a final invasion attempt against Taiwan, that the true test of air power was to be made over the straits.

Throughout this period, the Red Chinese were faced with unrest on the mainland. The invasion of Tibet, which took place during the Korean War, had not resolved the problem of Tibetan integration with the People's Republic of China. In 1954 and 1955, the PLAAF was engaged in an aerial campaign against the resistance forces in Tibet and ranged across the countryside to attack roving bands of Tibetan guerrilla fighters. In February, 1956, an underground revolt broke out in the ancient province, raging for five days before the

PLAAF was brought into action once more. The punishment was immediate and devastating. Tupolev TU-2 attack bombers, supported by other PLAAF aircraft, dumped about forty bombs on the insurgents in the first attack wave. The second attack wave put an end to the revolt, and load after load of bombs fell on the monasteries that harbored the Tibetan rebels. For months afterward, a lone PLAAF bomber circled the Tibetan capital at Lhasa as a constant reminder of Peking's military power.

The trend toward stabilization of Communist control over the mainland in the middle 1950's resulted in growing acceptance of the Chinese People's Republic on the part of other members of the Communist bloc. The next nation to aid Red China in the air was Poland, and in 1955 arrangements were completed for the export of a series of Polish SZD gliders to the PRC. A large glider center, under the direction of Polish engineers and other Polish advisers, was established. Gliding was adopted by the Chinese Communists as an inexpensive and efficient way to popularize flying throughout the country, much as it was used by the Japanese and Germans in the 1930's. The Central Gliding School at Anyang became the focal point for the new sport, and deliveries of the Polish gliders, accompanied by Polish instructors, took place in 1955 and 1956.

Produced by Szybowcowy Zaklad Doswiad Czalny, better known as the SZD Gliding Institute, the deliveries included fifty IS-3 ABC and ABC-A primary gliders; more than fifty Salamandra 53A primary gliders; and a substantial number of IS-4 Jastraab, SZD-8 Jaskolka, SZD-9 Bocian, and SZD-12 Mucha 100 sailplanes. They were quickly assigned to the PLAAF, where the primary gliders received the yellow-orange colors of Red Chinese trainers, and to civil gliding clubs and sports associations under rigid Communist management. Production licenses for the construction of SZD gliders in China were also obtained, and SZD designer J. Niespala was sent to China to establish a Red Chinese glider-design office and production facility at Tchan Tia-kuo.

The Chinese versions of the IS-3 ABC-A, SZD-8 Jaskolka, and SZD-12 Mucha 100 were virtually identical to the Polish originals, and they remained in production in the PRC long after manufacture of some of the models had ceased in Europe. The Tchan Tia-kuo design office, on the other hand, modified the Salamandra 53A design to remove the pod fuselage of the Polish version, producing an open-frame two-seat primary glider known as the China-Salamandra. In addition to the production of SZD types, the Tchan Tia-kuo facility also produced a catapult-launched single-seat primary glider of original Chinese design and in 1958 introduced the Jie-Fang Liberation 1 high-performance sailplane designed by Niespala before he left China. It first flew on May 10, 1958, and entered production with other glider types for PLAAF and commune sports groups.

The PLAAF's use of gliders was not restricted to training, and crack air force gliding teams participated in glider meets throughout the country, such as the exhibition matches at Wuhan, Harbin, and Peking in 1961. Tactical glider assault groups were also created in the PLA and flew large troop-carrying gliders, believed to be Russian-built Yakovlev YAK-14 transport gliders. Working closely with the assault teams transported to the Taiwan combat zone by Mil MI-4 helicopters, the glider troops had been trained for initial assault and support operations in the invasion.

While the early steps in the production of gliders were being taken at Tchan Tia-kuo, more impressive programs were under way in Manchuria. With Russian guidance, the Chinese Communists embarked on a twelve-year plan for the development of science and technology and established a number of Ministries of Machine Building to develop and manage the new production facilities being created. Jet aircraft, originally coming under the second Ministry of Machine Building responsible for the production of munitions and weapons, were of prime importance. Russian technicians, engineers, and industrial-management personnel convened in Shenyang at the site of the former Mukden plant of the Manshu Airplane Manufacturing Company, a large aircraft-production facility built in 1938 and operated by Japanese and Manchukuoan interests during World War II. When the Russians invaded Manchukuo early in August, 1945, the Mukden plant employed 4,965 people and produced aircraft engines and airframes for the Japanese Army. Manshu plants were also located at Kungchuling, where 1,462 people were employed to produce training planes, and at Harbin, where a work staff of 3,478 produced aircraft engines. The facilities were modern in every way, and within a matter of months, the Russians had removed about 95 per cent of the machine tools and equipment from the plants to the Soviet Union for their own use. The looting of the Manshu plants eliminated any possibility of immediate postwar aircraft production at the sites. However, the buildings and adjoining airfields remained intact, and many of the former Manshu employees, skilled in modern aircraft production, remained in the area to provide a labor pool for any reinstatement of production.

The creation of a new aircraft-production facility in Manchuria–reportedly the largest manufacturing plant built under Russian supervision in China in the mid-1950's–was one of the few cases in which the Russians found themselves rebuilding an industrial complex they themselves had destroyed. The National Aircraft Factory of the second Ministry of Machine Building was set up with its main plant at Shenyang under the direction of Russian production experts and a qualified staff of Chinese aircraft engineers who had studied in both America and Russia. Licenses were obtained to produce aircraft of Russian design as part of the first Chinese five-year plan, scheduled for the period between 1953 and 1957. This Soviet assistance was based on aid agreements signed in Moscow in

November, 1952, and October, 1954, the latter agreement providing China with the necessary production licenses, engineering drawings, and technical aid for the production of combat aircraft. The Chinese staff, supervised by Russian management and shop personnel, got the plant back into working order and in 1954 started modest production of the Russian Yakovlev YAK-18 primary trainer. The initial models were assembled in China from Russian components. The aircraft industry developed rapidly and the Chinese were soon producing both the required M-11PR engines under license and the complete aircraft. By the fall of 1956, the Chinese Communists had produced their first combat aircraft, a Shenyang MiG-17 fighter assembled from components and assemblies supplied by the Soviets. At the end of the year, the PLAAF proudly took delivery of its initial Shenyang fighters, each with the character for China on the nose followed by a four-digit identification. Six months later, the production rate was fifteen aircraft per month with Russian-built VK-1 engines. By the middle of 1959, production of Shenyang MiG-17 fighters with Chinese versions of the Russian engines had reached the rate of twenty-five aircraft per month. The Chinese power plants had been developed, and production begun, under the supervision of the Institute of Mechanics of the Chinese Academy of Sciences, an establishment set up early in 1956 under the direction of Dr. Chien Hsueh-shen to study jet propulsion and produce jet engines in Communist China as part of the twelve-year plan for self-sufficient technology.

The Communist Chinese need for advance fighter trainers for the expanding PLAAF was also met by the National Aircraft Factory, and a Chinese version of the MiG-15UTI two-seat advanced trainer was soon produced on parallel production lines with the Shenyang MiG-17.

As the Chinese gained confidence in their aircraft-construction capability, additional plane types were added to the expanding production lines. Up to this point, all production had been for military purposes, but in December, 1957, the Chinese Communists completed their first civil aircraft, a license-built version of the Russian Antonov AN-2 biplane. Once again, the initial engines were Russian-made and the airframes were built in China. Later, the engines would also be produced in China.

The first Shenyang AN-2 flew early in 1958. The type, produced in China as the Fong Shou No. 2, was soon in great demand for civil passenger transport and agricultural duties as well as for training and transport use in the PLAAF. Production of this model doubled in one year, and a substantial number were built for the Special Flight Group, a branch of the civil-aviation authority serving agricultural and industrial needs, and for geological-survey agencies and sports groups. The Fong Shou No. 2 was soon powered by the 1,000 hp Chinese Model 670-13 engine, freeing production from any dependence on the Russians. Production also began on a Chinese version of the AN-2S, a modified AN-2 with a motor-driven pump to distribute liquid and dry fertilizer or insecticide from the air. These aircraft were made available to various agricultural communes and farming communities to help in pest control and forestry work.

In their news media the Chinese boasted of their ability to produce aircraft and, more often than not, claimed that their new Shenyang-produced fighters, trainers, transports, and utility types were of original Chinese design. This, of course, was not true, for the Chinese were still almost totally dependent upon Russian aid and generosity. According to Dr. Chien Hsueh-shen, as revealed in a published interview in October, 1956, the National Aircraft Factory was producing MiG-17 fighters at a time when the facility had no design office and none, in fact, was even in existence at the Institute of Mechanics. It was Dr. Chien's belief that China's ability to design and produce her own aircraft remained only a hope for the future, and he stated that China had no immediate plans to build aircraft of original design. But during this period, things were changing in the Soviet Union. The death of Stalin in 1953, coinciding with the end of the Korean War, led to a series of power struggles in the U.S.S.R. The emergence of Nikita Khrushchev as the dominant Russian leader would have far-reaching effects on the Sino-Soviet alliance. Whereas Stalin had regarded Red China as an extension of his own ambitions in the Orient—and a risk that would pay handsomely in the long run—the new Russian hierarchy viewed the Chinese mainland in terms of its trade opportunities. Soviet aid to China had a monetary value, and the Russians kept a running record of the Red Chinese debt. During the so-called honeymoon period in the late 1950's, this did not affect the Sino-Soviet relationship; but as ideological differences arose in 1959 and 1960, it was to contribute to the split between the two Communist states.

The year 1958 was the high-water mark for Red China in the air. The strength of the PLAAF was at an all-time peak, domestic aircraft production was getting into full swing, and Red Chinese credit was established. As if to belie Dr. Chien's pessimistic statements of 1956, a virtual explosion of new air-industry developments took place. Production of the Russian Mil MI-4 helicopter was undertaken by the Shenyang complex in 1958, and the first civil Chinese version, the Whirlwind, was announced in 1959. The Chinese MI-4 entered PLAAF military service, joining the Russian-built prototypes, and a substantial number of the civil models were built for the Special Flight Group. The PLAAF helicopters often aided in civil work after national disasters, such as earthquakes or droughts. Shenyang MI-4's were also put on view at air shows to make the populace more air conscious, and one of the first models was exhibited at the National Defense Sports Association Aviation Show in Peking on August 10, 1958.

The Russian Yakovlev YAK-12 general-purpose monoplane was produced in China as the Chinko No. 1. The Chinese prototype, it was reported, was built in "seventy-five days" by skilled workers, professors, and students at the Shenyang aviation school; it ultimately entered production in Manchuria as a trainer, liaison, air-ambulance, communications, passenger transport, agricultural, and general-utility type for the PLAAF and the Special Flight Group. Dramatic announcements telling how long it took to produce original Chinese aircraft, and the number of "workers, professors, and students" involved became quite common as new Russian aircraft with their Chinese modifications were adopted for production.

The Red Chinese concept of technical education was to combine theory with practice, leading to the establishment of aeronautical engineering schools on the same sites as the aircraft-producing factories. The engineering students studied in class with texts and teaching aids, and then actually worked in the assembly shops. This dual knowledge, of basic design and of production requirements, was to pay great dividends in later years after the Russians pulled out of China and left the Chinese Communist aircraft industry on its own.

A twin-float version of the YAK-12 was developed and produced for agricultural work at the Feilung Machine Building Works in Shanghai in the fall of 1958 in "forty-eight days" and entered production as the Hiryu ("Flying Dragon"). A highly modified three-passenger version for crop dusting with the fuselage redesigned to provide a large-capacity chemical hopper was built at the Harbin Aeronautical Engineering Works in Manchuria; it was designated as the Hai Lun-kiang No. 1 and first flew on December 16, 1958. Another YAK-12 variation; the Yenan No. 1, for sports and agricultural work, was built at the Northwestern Technological University at Sian in Shensi province, in 1959, making the YAK-12 possibly the most prolific basic design in the Chinese aircraft industry.

The Chinese also undertook the production of multiengine

aircraft, although the early examples were of simple design and construction. Assembly of the prototype Bei-Jing No. 1 transport, an original Chinese design based on the Russian Yakovlev YAK-16, was in progress during the summer of 1958 at the Peking Aeronautical Engineering College, Red China's leading specialized aeronautical engineering school with its own adjoining manufacturing complex. At the aircraft's press introduction on October 1, 1958, the ninth anniversary of the founding of the People's Republic of China, it was claimed that the Bei-Jing No. 1 had been "designed and built in 100 days." Even the most advanced aircraft-producing nations would find it difficult to match this so-called record, and it was obvious that the Chinese had relied heavily on the design of the larger Russian type. Nevertheless, the Chinese Communists could rightfully be proud of their achievement, no matter how much time it actually had taken, although the Red Chinese records were wrong on another count. The plane was said to be the first Chinese transport, but the Naval Air Establishment (NAE) facility at Foochow had designed and built a transport in the mid-1920's, and a series of modern low-wing twin-engine monoplane transports were under development by the Nationalists in 1948 as the Chung-Jun series (this was later dropped as the civil war swept over China, and the prototypes may have fallen into Communist Chinese hands). The fact that these transport aircraft were designed and produced under non-Communist governments seems to have precluded their existence in Red Chinese eyes.

The prototype Bei-Jing No. 1 was turned over to the Chinese national airline for evaluation as a feeder-line transport, and production models entered PLAAF service as light transports, multiengine trainers, and communications types. More conclusive examples of original Red Chinese work also showed up in this period, although the work may have been directed by Russian engineers. A scaled-down version of the Russian Antonov AN-14 light transport was built in prototype form at the Peking Aeronautical Engineering College in 1959 in "sixty-eight days." The extensively modified Chinese version was designed for four passengers, whereas the larger Russian model had been built for eight. The Power plants of the Chinese version were also built in China and the small utility transport entered production as the Peking Sha-Tu, or Capital No. 1. In addition, an apparently completely original monoplane was designed to meet the same requirements as the Shenyang Fong Shou No. 2 biplane then produced under a Russian license. Identified as the Red Banner No. 1, this new utility transport was built in prototype form at the Peking Aeronautical Engineering College late in 1958.

The Chinese turned to Czechoslovakia for an aircraft to be used as a feeder-line and light communications type. The aircraft was the highly successful Super-Aero 45, then widely used in the Soviet Union on short-range Aeroflot routes and in other nations as well. A modified Chinese version was built at the Harbin Aeronautical Engineering Works in Manchuria; the first example was completed in "eighty-two days" and was proclaimed another Chinese original. It remained in production at Harbin as the Sungari No. 1 (the name of a river in Manchuria), and deliveries were first made to the Red Chinese national airline in 1959 to join the Shenyang passenger version of the AN-2 biplane on short routes.

Civil aviation in Red China was created virtually overnight when the pilots and crews of nine Curtiss C-46 and Douglas C-47 civil transports of the China National Aviation Corporation (CNAC), one of the two major Nationalist civil airlines, defected to the Communists on November 9, 1949, as the Nationalists were being chased off the mainland. Seventy-one of the remaining CNAC and Central Air Transport Corporation (CATC) aircraft were interned at Kai Tak Airfield and in the harbor at Hong Kong, pending a decision regarding their legal ownership. On February 23, 1950, the Hong Kong Supreme Court ruled that the aircraft rightfully belonged to the

People's Republic of China, thereby dealing a deathblow to the Nationalist airlines while giving the Communists everything they needed to start operations on the mainland. Within a few days, Communist crews descended upon Hong Kong to paint out the old Nationalist XT registrations and apply the People's Republic of China flag markings on the tail, but they never received the aircraft. In July, 1952, the ruling was reversed, and the impounded aircraft reverted to American ownership.

The Red Chinese had already established a structure for their national airline at the time of the take-over. Airline operations, both domestic and foreign, came under the jurisdiction of the Civil Aviation Administration of China (CAAC), with offices in Peking. Routes were laid out between major Chinese cities, and later in the year a route was set up between Harbin, in Manchuria, and the Siberian border town of Chita in the Soviet Union. Most of the traffic between China and Russia, however, was handled by Russian Aeroflot transports operating direct to Peking from the Soviet Union.

With the coming of the Korean War and increased Soviet aid, additional transports were acquired from Russia to augment and replace the aging Nationalist airliners. On New Year's Day, 1955, another direct link to Russia was opened from Sinkiang Province to Alma-Ata, where the line connected with an Aeroflot link to Irkutsk and central Russia. The aircraft flown by the CAAC were Russian-built Lisunov LI-2 transports, license-produced versions of the Douglas DC-3. Lisunov transports were also used on the domestic routes, and they were soon joined by Russian-supplied Ilyushin IL-14M transports, twenty of which were obtained in 1956.

The first scheduled route to a non-Communist state was initiated on April 11, 1956, when a CAAC Ilyushin IL-14M left Kunming at 7:30 A.M. and arrived at Rangoon, Burma, at noon. The IL-14M was the first modern CAAC transport and was assigned to high-frequency passenger routes. CAAC advertising and timetables illustrated the Ilyushin transports, and these aircraft were licensed in three-digit 600 series registrations for civil use. New routes were opened, including an IL-14M link between Peking and Sining in Tsinghai Province in January, 1957.

Substantial increases were made in the CAAC civil air fleet in 1958 as part of Mao Tse-tung's Great Leap Forward policy, beginning with the import of a number of twenty-six-passenger IL-14P transports produced under license in East Germany at VEB Flugzeugwerke, Dresden. As the first CAAC type available in quantity, the IL-14P became standard equipment and began to replace the remaining LI-2, C-46, and C-47 types still in use. The growing need for these transports led to a repeat order, and a new batch of German VEB transports was flown via Moscow to Peking in the fall of 1959.

The other great expansion during this period was in the growing number of feeder lines opened throughout the country. The stated objective was to bring all of China together by means of connecting air links. The new feeder lines were widely heralded in the Chinese press, and ceremonies were held at their inauguration. Shenyang AN-2 transports were the most widely used because of their ability to land at difficult locations. A typical example was the route opened on July 11, 1958, between Taiyüan and Changchih in southeastern Shansi. The overland route, across rough terrain, took four days; the CAAC flight covered the distance in two hours.

On high-density feeder lines, such as that between Canton and Hankow, the Shenyang AN-2 biplane transports were joined by Czech Super-Aero 45 monoplanes. Other AN-2 and Super-Aero 45 routes connected Kunming, Chengtu, and Kweiyang in south-central China, as well as the Chinese nuclear test site in Sinkiang, with other cities in northwestern and north-central China. These routes—and the aircraft—remain in service at the present time.

In 1959, the CAAC claimed that practically all the "old-fashioned U.S. planes" had been replaced with modern Ilyushin and Chinese-built transports. This statement was not wholly true. The older planes were certainly replaced on passenger routes, but many of them remained in service as freight and cargo transports. Special duties included the flying of newspaper mats of the *People's Daily* from Peking to outlying districts to make daily national distribution of the paper possible. Lisunov LI-2 and Douglas C-47 transports were also used to ship livestock—pigs, sheep, cattle, and even pedigreed horses—across the country. These aircraft are still in use and constitute a ready fleet of transports for military operations, although their load limitations and short-range characteristics prevent their classification as modern transports.

While CAAC services were being expanded in China, steps were also being taken in Russia to upgrade major CAAC routes domestically and between Peking and the Soviet Union. Chinese pilots, crewmen, and mechanics were sent to Moscow late in 1958 to begin training on the four-engine turboprop Ilyushin IL-18. Red China was one of the first export customers for the IL-18, and upon completion of training in 1959, the new crews returned to China with an original order of five IL-18 transports to open routes between Canton and Peking; later, CAAC routes CA-135 and CA-136 between Peking and Irkutsk in Russia were added, the latter connecting with an Aeroflot link direct to Moscow.

While aircraft have been added since the 1958 build-up, the Communist civil air fleet has not been significantly modernized, although attempts have been made to improve CAAC equipment and service. The Great Leap Forward of the late 1950's lost its momentum, and the ideological split that took place between Communist China and the Soviet Union in the early 1960's brought many ambitious civil-aviation plans to a grinding halt.

The Ivchenko AI-20 engines of the Soviet IL-18 transports

frequently broke down and had a poor service life. In 1963, the engines were changed, and five new IL-18 transports ordered from Russia had improved power plants. In spite of its poor performance, the IL-18's became the pride of Red China, and they were used for international good-will flights to central and southeastern Asia, the Middle East, and Africa, carrying Chinese Communist dignitaries to various international conferences and meetings. At least one was turned over to the PLAAF as a VIP military transport. When Chairman Liu Shao-chi, regarded until the summer of 1966 as the obvious successor to Mao Tse-tung, flew to Indonesia, Burma, Cambodia, and North Viet-Nam on good-will missions in the spring of 1963, the aircraft used for the flights was a former IL-18 civil transport with PLAAF military markings. Premier Chou En-lai's flight to Moscow in early November, 1964, for the forty-seventh anniversary of the October Revolution was also made in a military IL-18—possibly the same aircraft.

These flights and Red Chinese negotiations with other countries are symptomatic of China's eagerness to establish international flag routes around the world that connect directly to the Chinese mainland and Peking. Major Chinese airports, such as those at Canton, Shanghai, and Peking, have been completely modernized for all-weather IL-18 and jet operations. Yet as late as the summer of 1967, the Chinese had yet to put jets on regular service on CAAC routes. An air agreement between China and Pakistan in June, 1963, allowed both countries to have routes from Canton in China to Dacca, the capital of East Pakistan, but the Chinese have not yet exercised this option. Pakistan International Airlines (PIA), on the other hand, now regularly flies into Red China with Boeing 720B jets, giving the Chinese the opportunity to observe these transports in operation. Language difficulties led to the use of English in the control towers at Canton and Shanghai for communication with the PIA jet pilots, a fact that must certainly trouble the Chinese.

The recognition of Red China as the "sole legal government of China" by France in January, 1964, and good relations with Cuba during the early 1960's led the Chinese Communists to think about long-range air links with Paris and Havana, and aircraft for the planned routes were eagerly sought. In January, 1963, while CAAC representatives were in Great Britain to negotiate the purchase of Vickers Viscount transports for internal Chinese routes, they seriously considered purchasing a number of former BOAC Bristol Britannia 102 airliners for international use. The British had removed these large six-year-old turboprop transports, used on overseas flights until replaced by jets, from the Peking embargo list and—in spite of protests from India as a result of the border fighting between Red China and India in 1962 and 1963—were ready to sell them to the Communists for about $420,000 each. CAAC pilots flew an example on a demonstration flight in England, but after due consideration, Peking finally turned down the offer.

It is obvious that the Red Chinese rejection of the Britannia offer was based on the obsolescence of the aircraft, for the Chinese then made a tentative counterrequest to purchase more advanced Vickers VC-10 long-range jets, among the most modern jet transport aircraft in the world, for the planned Chinese route to Havana, as well as the smaller British BAC-111 short-range jets for Chinese internal routes. The Chinese had carefully evaluated both aircraft at the British aircraft exhibit at Farnborough in 1961 and knew what they would be buying. The British, fearful of the American reaction to such a sale, quietly rejected the suggestion before serious negotiations could get started. British interest in the Red Chinese market remains high, however, and Hawker-Siddeley displayed models of its DH-121 Trident transport at the British traders' exhibition at Peking in November, 1964, and was ready to discuss the possible sale of these medium-range jets to China. We have not heard the last of this yet, and it can be expected that the Chinese will purchase modern jets in the near future. Such a purchase would greatly expand Chinese airlift capabilities and would therefore have important military implications.

The PRC also turned to France for long-range jets and in the fall of 1963 seriously considered the purchase of SUD-Caravelle SE-210 transports for use on a route to Paris. Once again, they ran afoul of American interests, for the power plants would either be GE's of American origin or Rolls-Royce engines from England. By the spring of 1964, the deal had fallen through for one reason or another. At the end of the year, the Chinese placed an order with the Soviet Union, by now their sworn enemy, for an additional five IL-18 transports, bringing the CAAC total to fifteen. These may possibly be used on the Peking-Paris route, but the service has yet to be started. An Air France route, however, has been in operation since September 19, 1966, so the air link now exists, although the route is not a direct one. Landing rights at Peking have not been granted to Air France, so the service is between Paris and Shanghai on a once-a-week basis via American-built Boeing 707-328 jet transports. Leaving Paris on Monday mornings at 11 A.M., the Air France jet arrives at Hungjao Airport in Shanghai at 2:40 P.M. the next day, making stops at Athens, Cairo, Karachi, and Pnompenh on the way. Connections from Shanghai to Peking are made on the high-frequency route flown daily between the two cities by CAAC IL-18 turboprop transports. CAAC flight CA-054 leaves Hungjao every Tuesday afternoon at 4:50 P.M. and is scheduled to arrive in Peking at 7:10 P.M., so a Chinese diplomatic passenger or a French businessman can travel halfway around the world from Paris to Peking in a little over thirty-two hours. The Air France jet starts back for Paris the same night it arrives, so a returning visitor would leave Peking on a Tuesday afternoon on the daily 1 P.M. flight to Shanghai, arriving at Hungjao at 3 P.M. The Air France Boeing leaves Shanghai at 5:30 P.M. and arrives at Orly Airport in Paris on Wednesday at 9:30 P.M.

The international air routes of the CAAC, while they have expanded modestly in the past few years, remain short-range routes to closely neighboring countries. The CAAC now flies to Rangoon in Burma, Vientiane in Laos, Hanoi in North Viet-Nam, Pnompenh in Cambodia, and Pyongyang in North Korea. The routes to the Soviet Union and Ulan Bator in the Mongolian People's Republic have been cut as a result of the Sino-Soviet split, although a Soviet Aeroflot link still connects at Harbin in Manchuria and at Peking, as does a Mongolian feeder route from Ulan Bator. Plans for CAAC flights to Indonesia were shelved when the Communist take-over of that country was thwarted in September, 1965, and the shrinking CAAC routes reflected the growing isolation of Red China on the world stage. Most of the present routes are handled by the now obsolete IL-14M and IL-14P transports obtained in the late 1950's.

Some effort at upgrading domestic routes has been made in the past few years, although the improvements in no way bring CAAC service in line with that of other major nations. In January, 1961, representatives of the Civil Aviation Administration of China and the National Trading Corporation initiated discussions with Vickers interests in England to purchase six short-range Vickers Viscount 843 turboprop transports. In spite of American attempts to block the sale, the first of these aircraft were ready for delivery in June, 1963, and arrived in Hong Kong on July 6. They were put into service on domestic routes as soon as they arrived in China and constitute the most modern airliners in CAAC service. Their arrival was none too soon, as they were sorely needed. Air Commodore M. Nur Khan, managing director of Pakistan International Airlines, and his staff were shocked at the time of their signing of the air agreement with Red China in June, 1963, when they realized that their CAAC flights would be on obsolete IL-14P airliners. Other visitors to Red China are equally taken aback, and the over-all impression is that the Chinese air industry is behind the times. On the other hand, the sale of two British Handley Page HPR7 Herald transports in the Fairchild Friendship class in June, 1965, has been reported, and apparently the Chinese have an option to buy more of these aircraft. But even at best, the CAAC civil passenger-carrying fleet of six Viscounts, two Heralds, fifteen IL-18 long-range transports, as well as more than a hundred Russian and German IL-14 types, various models of the Shenyang AN-2, Bei-jing No. 1, and the Super-Aero 45, and another hundred cargo and utility aircraft leaves much to be desired for a nation with the ambitions of the People's Republic of China.

The People's Liberation Army Air Force has also been faced with the problem of creeping obsolescence, but there are signs that this cancer has been blocked and that the PLAAF has turned the corner toward a modern air arm. Red China's current troubles in the air began in 1958, at the very time when the PLAAF was confident that it was ready to take on the Nationalists over the Quemoy straits. After the Korean War, with Russian aid the PLAAF had created a well-organized, disciplined, and aggressive air arm, and production of the Shenyang MiG-17 fighters had gone into full swing. Mao Tse-tung's militant threats to invade Taiwan had world opinion almost convinced that the Nationalists would be wise to submit to Communist domination. Only the U.S. Seventh Fleet and a strengthened Nationalist military force stood in the way. The battle for Quemoy was close at hand.

Opposing Chinese jet fighters had clashed over the straits in the past, with sporadic victories going to the Nationalists, starting with an encounter on May 11, 1954, when the most advanced fighter in the PLAAF was still the Russian-built MiG-15bis of Korean War vintage. These off-and-on engagements gave no indication that the Communists were making a calculated effort to drive the Nationalist Air Force out of the skies. But the battle for Quemoy was different. The first sign of the impending conflict was increased Communist air activity over the straits; flights of Red Chinese MiG-17 fighters often came within minutes of the Nationalist air bases on Taiwan. On July 29, 1958, four PLAAF MiG-17 fighters, led by Company Commander Chao Teh-an, tangled with four Nationalist fighters over the straits, and in two minutes, according to the Red Chinese claims, the Communists downed one Nationalist fighter and damaged another. But on August 14, in an engagement over Matsu, the Nationalists shot down three MiG-17's and had no losses. Less than a month later, on September 8, the Nationalists announced that they had destroyed seven MiG-17 fighters and had damaged two just off the coast of mainland China. Again no Nationalist losses were reported. At this point, the Communists cried foul, and the fighting was intensified. Although both sides claimed victory over the straits, with the Communists even declaring that PLAAF pilot Pao Shou-ken had shot down enough Nationalist planes to be called an ace, the facts indicate that the battle was a clear-cut Nationalist triumph.

Nationalist Air Force morale soared during the battle, and as the F-86F Sabre jets returned from combat with their reports and confirming gun-camera records, the scope of the victory became apparent. On September 18, Nationalist pilots downed another five MiG-17's; on September 24, they destroyed ten more and damaged up to six. On October 10, the Nationalists announced the loss of an F-86F Sabre in an engagement off the mainland coast that had cost the PLAAF five more MiG-17 fighters as well as two damaged planes that had limped back to base. At this point, the Communists disengaged. Defeated in the air, and faced with a Russian refusal to provide the needed military equipment for the planned invasion attempt, the Chinese Communists called off their projected assault on Taiwan. It wasn't until July 5, 1959, that

Taiwan claimed another victory—the destruction of five MiG's over Matsu at no loss to the Nationalists.

The importance of these victories lay in the fact that the Nationalists had defeated an enemy that outnumbered them by a wide margin, and they had done so on their own. Although the Nationalists were armed with Sidewinder missiles late in the fighting, most of the early victories had been obtained with conventional armament. The fact remained that the Nationalists had chased the PLAAF out of the skies over the Quemoy straits so completely that to this day the Red Chinese only rarely dare to fly over the offshore islands, then quickly dart back to the mainland. The final tally for the 1958 fighting, according to the Nationalists at the time, was the destruction of thirty-one Communist MiG's as compared with the loss of two Nationalist fighters, a kill ratio of almost 16:1.

By the end of 1965, combat over the straits had become such ancient history that the Nationalists were completely confident of their control of the air and almost forgot about the possibility of Red Chinese reprisal. The last encounter had taken place on February 16, 1961, when the Nationalists shot down a single MiG-17 in a brief engagement with no loss to themselves. With this kill, the Nationalist claims put the Communist losses over the straits between May 11, 1954, and February 16, 1961, at two MiG-15 and forty-one MiG-17 fighters, plus two probables and ten possibles, with the loss of only two Nationalist fighters (Taiwan now claims only a single loss for the period). The air over the straits became a Nationalist sanctuary, and Convair PBY-5A and Grumman HU-16 Albatross courier planes flew shuttle flights between Taiwan and the offshore islands for years without Red interference. But on January 10, 1966, the Nationalists were reminded of the ever-present threat of the PLAAF. On the previous day, three men out of the ten-man crew of a Red Chinese landing craft had made it safely to Matsu in a dramatic defection from the mainland. The Nationalist press loudly proclaimed its welcome to the Red defectors and announced that they would arrive on Taiwan the next day. The next afternoon, when the men were picked up by a Nationalist HU-16 for the flight to Taiwan, three Red Chinese MiG-17 fighters flashed out of the sky, shot down the Albatross as a warning to future defectors, and just as quickly returned to the mainland. There were no survivors.

Following their defeat over the Quemoy straits in 1958, the Red Chinese were faced with another problem, that of the increasing Nationalist overflights of the mainland. The Nationalists had long been flying over the Communist coastal areas, as well as southern and central China, for propaganda and reconnaissance purposes and only rarely encountered PLAAF aircraft. This overflight program had started as soon as Chiang Kai-shek's government had reached the safety of Taiwan. During the famine of 1950, unarmed Nationalist Douglas C-47 and Curtiss C-46 transports from Taiwan dropped bags of rice and other foodstuffs over hard-hit areas on the mainland. Stenciled messages on the bags reminded the receivers that the Nationalist cause was still very much alive. As the Korean War years rolled on and as Red China grew in strength and importance, the complexion of the Nationalist overflight program changed dramatically. These flights became more dangerous as the Communists built up their internal air defenses. Food and propaganda leaflet drops continued to be made, but a number of guerrilla and intelligence drops were added when the need grew for accurate, up-to-date military-intelligence data—to be evaluated by the United States and other Western nations—in order to pierce the Bamboo Curtain. Aerial reconnaissance with high-performance aircraft provided an efficient way for the West to keep tabs on Red Chinese military and atomic progress, and the Nationalists accepted the responsibility of flying specially equipped reconnaissance aircraft supplied by the United States. In the early days of the reconnaissance overflight program, the Nationalists used Lockheed RT-33, Republic RF-84, and Douglas RB-66 aircraft. These were later joined by McDonnell RF-101, Martin RB-57D, and Lockheed U-2 types. The overflight crews photographed mainland installations at Peking, Tientsin, Taiyüan, Lanchow, Shenyang, Sian, Nanking, Wuhan, Shanghai, and other locations, as well as the growing Chinese Communist nuclear complexes at Lob Nor.

The PLAAF units assigned to homeland defense were primarily equipped with Russian and Shenyang MiG-17 fighters, although MiG-15's were also in widespread use. They were found to be ineffective as high-altitude interceptors, and the PLAAF was continually frustrated by its inability to knock down the high-flying Nationalist reconnaissance planes. In 1959, complaints to Russia led to the delivery of a number of Soviet Mikoyan MiG-19 fighters equipped with radar gunsights, two powerful jet engines, and air-to-air missiles. Antiaircraft ground-to-air missiles were also delivered, accompanied by Russian technicians to instruct the Red Chinese in their use.

On October 7, 1959, the PLAAF shot down an RB-57D over northern China, and the battle for mainland air control was under way. It wasn't until August 2, 1961, after a frustrating summer during which Nationalist planes had overflown Peking without interference, that another reconnaissance plane was brought down. This time it was an RF-101 taking photographs over Fukien Province, and the Nationalist pilot was captured alive. On November 6, 1961, an explosive incident almost resulted when the Chinese Communists shot down a U.S. Navy P2V–7 Neptune reconnaissance plane, claiming that it had overflown northeastern China. An angry exchange of notes took place between the United States and Red China through intermediaries in the absence of any diplomatic exchange between the two countries.

At about this time, Lockheed U-2 reconnaissance planes, the first of which had been delivered to the Nationalists in

1960, showed up over the mainland, and the Red Chinese were faced with a more difficult target. In 1963, Nationalist U-2 flights revealed that the PLAAF had fallen on hard times, that its strength was deteriorating. Observation disclosed that only a small number of late-model MiG-19 fighters were in service. Obviously, the withdrawal of Russian aid in 1960 had had a profound effect on the line strength of the PLAAF. Cannibalized aircraft and growing scrap piles indicated that the Red Chinese were confronted with a crisis in maintenance, one that is only now being overcome. Of prime importance were the U-2 intelligence reports on Red China's nuclear progress. These flights provided a running record of the activity at Lob Nor, and it was therefore possible to anticipate the PRC's nuclear tests. As a result, the world was ready for both the first Red Chinese atomic explosion, on October 16, 1964, and the second, on May 14, 1965.

The first Nationalist U-2 was shot down over Nanchang, Kiangsi Province, in south-central China on September 9, 1962. Much of the aircraft remained intact, but the pilot, Colonel Chen Wai-sheng, the same Nationalist officer who had spotted the first Chinese MiG-17 while flying over Shanghai six years before, was killed. A little more than a year later, on November 1, 1963, another was brought down over Hua Tung, near Shanghai and the coastline. The third, which the Nationalists claimed to be a loss caused by an in-flight accident, came down over the mainland on July 7, 1964, and the fourth was brought down in January, 1965. Obviously, the Red Chinese had developed the capability to bring down even the highest-flying reconnaissance aircraft in use, and the four wrecked U-2 aircraft were proudly displayed by the Communists at the Chinese People's Revolutionary Military Museum in Peking in April, 1965.

Even low-flying reconnaissance types have been caught in the air. In the spring of 1965, pilot Kao Chang-chi of a PLAAF regiment stationed along the China coast was credited with bringing down an RF-101 of the Nationalist Air Force in a dirty-weather interception, and pilot Tung Hsiao-hai, company commander of the same unit, was credited with shooting down a high-altitude American aircraft in April, 1966. This regiment, equipped with Shenyang MiG-17 fighters as late as summer, 1966, reputedly shot down or damaged thirteen Nationalist and American planes from the time of its service over the Quemoy straits in 1958 up to 1966—not an unreasonable figure, but certainly an exceptional one when compared with the record of many other fighter regiments in the PLAAF.

The PLAAF's ability to defend its own skies, although long in coming, reflects a growing capability to protect Red Chinese cities and industrial sites against bombing attacks. The overflight threat resulted in the installation of ground-to-air rockets around Peking and Shanghai; later, as more rockets became available, they were placed around other prime sites. PLAAF interceptor installations—along the east coast, facing Taiwan; in the south at Nanning, in Kwangsi; on the island of Hainan, in answer to the American air build-up in Viet-Nam; and along the Russian border—have been enlarged and strengthened, and the basic philosophy behind the PLAAF has increasingly become defensive by force of circumstance if not by capability.

The Chinese Nationalists have long requested moth-balled American Boeing B-47 Stratojet strategic bombers for use against Red China. Following the border war between Red China and India in 1962 and 1963, in which few aircraft took part, the Indians also made tentative bids for B-47 bombers so that they could avenge themselves on Red China. The B-47's would have given both Nationalist China and India a far better offensive air arm than that maintained by Red China. The U.S. Government has refrained from meeting these requests, for such a move could trigger a third world war. The possibility of all-out war was inherent in the Sino-Indian conflict in the summer of 1963. More than 1,000 Red Chinese fighters and a regiment of Tupolev TU-4 bombers were poised and ready for use against the Indians. The crisis was averted only because of the Red Chinese military victory in the border fighting and the resulting Indian humiliation. But control of the air along the borders of Red China remains a burning question, one that has been contested more than a few times during the fighting in Viet-Nam. The resulting air incidents between the United States and Communist China have been a source of considerable tension.

CHAPTER III

SELF-RELIANCE AND THE FUTURE 1965–70

As soon as the USAF Douglas RB-66 Destroyer radar-jamming jet was attacked by the four MiG-17 fighters over North Viet-Nam, its covering fighter patrol of three F-4C Phantom jets closed in for a counterattack. In moments, one of the Communist fighters exploded with a Sidewinder missile up its tail pipe, while a second was damaged and limped north. The date was May 12, 1966; the time, shortly after 4 P.M.; the location, about 115 miles northeast of Hanoi and an estimated 50 miles south of the Chinese border; and the incident, explosive—for the fighters were Chinese.

The Red Chinese reaction was immediate and confusing. According to Radio Peking, the air battle took place northeast of Makwan, over Yunnan Province, in the People's Republic of China, twenty miles north of the Vietnamese border, and the action consisted of a surprise attack by five American aircraft on a single Chinese plane of the People's Liberation

Army Air Force on a training flight. As evidence, the Communists produced pieces of a Sidewinder missile and seven American auxiliary fuel tanks reportedly dropped in the Makwan area. The Chinese stated that the American aerial intrusion was a deliberate act of war and reported that the U.S. fighters had fled as soon as they were counterattacked by a flight of PLAAF interceptors.

The full truth may never be known, but the significant fact remains that the Chinese Communists admitted their first fighter loss to American aircraft since the bombing of North Viet-Nam began, in February, 1965. Although the American pilots later stated that it was inconceivable that the fight took place over Chinese territory, it is all too clear that Red Chinese and U.S. aircraft are in close combat proximity and daily run the risk of facing each other eyeball to eyeball along Red China's common border with North Viet-Nam.

One fact is known: American aircraft have, without question, violated Red Chinese airspace; on more than one occasion, aerial combat has resulted. One look at a map of North Viet-Nam and the Gulf of Tonkin will explain some of the reasons why, for the speed of modern jets, coupled with the curving borders of Red China, makes it difficult for American pilots to follow their rigid instructions to avoid Chinese territory. Consequently, there has been something approaching an unofficial and undeclared air war between the two powers, and, to date, Red China appears to be ahead. The first encounter took place in August, 1964, when thirty American jets took off from South Viet-Nam to meet a flight of Chinese fighters from Hainan. The Chinese jets went into a holding pattern short of the coast, and the planes never met. On April 9, 1965, the Chinese Communists claimed that eight U.S. Navy F-4B Phantom jets flew over Hainan in two waves and that when PLAAF interceptors engaged them in a dogfight, one of the American planes mistakenly downed another with a Sidewinder missile. The American version of the engagement differed considerably, for the Navy stated that four of its Phantoms, on anti-MiG patrol, were attacked by an equal number of unpainted and unidentified MiG-17 fighters thirty-five miles south of Hainan. The battle reportedly ended when the MiG's raced for cloud cover, with one streaking flames. No American loss was reported.

Other clashes were more conclusive. On September 20, 1965, a People's Liberation Army Naval Air Force fighter downed a USAF F-104 Starfighter over the Hoihow area of Hainan, and the American pilot was captured and interned. In April, 1966, PLAAF pilot Tung Hsiao-hai shot down an intruding Douglas A-3D Skywarrior over the Luichow Peninsula, which juts out from mainland China east of Hanoi into the Gulf of Tonkin, and at least one other American plane was brought down over Kwangsi Province. In the beginning, U.S. policy was to ignore or deny Chinese Communist claims regarding the intrusion of Red China's airspace by American aircraft. The result was a propaganda advantage for Peking. The People's Republic of China was able to cry out in righteous indignation over the supposed American invasion of its skies with little or no counterreaction by the U.S. Government. The Foreign Ministry in Peking periodically issued "serious warnings" to the United States charging provocation by American aircraft, although it was obvious that the announcements were also meant to inform the larger world audience. By the start of 1967, more than 400 of these open complaints had been made in print or over Radio Peking, many of them having the ring of truth.

The PRC's attempt to maintain the integrity of its airspace, a defensive policy long pursued by China, at last received reluctant official acceptance in Washington. Steps were taken by the United States to counter Peking's claims that American aircraft were intentionally sent across the border to provoke the Chinese into all-out war. Washington subsequently instructed the USAF and naval carrier forces to take all necessary precautions to avoid Chinese Communist territory by remaining twenty-five to thirty miles away from the northern border of North Viet-Nam and the coastal areas of mainland China and the island of Hainan. Similar restraint has been maintained by the Chinese Communists, once again to their advantage. While the United States sees Red China as an aggressive giant intimidating its neighbors, few complaints regarding overflights of bordering territory by PLAAF aircraft ever reach the stage of excitement that the Communists are able to generate over an American intrusion. As much as the United States would like to have a clear case of aerial aggression to slap against the air forces of Red China, it just hasn't materialized. This leaves Washington with little to say to counter PRC complaints, such as the "425th serious warning" of February 22, 1967, when the Red Chinese charged that two American aircraft had strafed and damaged two Chinese fishing boats in the Gulf of Tonkin two days earlier.

These reports indicate that the Red Chinese are able to track and report many real or imagined intruders and to exercise a degree of judgment regarding the interception of the aircraft. The volume of complaints and the low number of actual interceptions, under conditions that would appear to be advantageous to the Chinese, add an air of mystery to the whole situation. The easiest conclusion to draw is that the Chinese are incapable of defending their territory against foreign aircraft and that attempted interceptions are usually unsuccessful. This has certainly been true in some cases, but the reverse may actually be the case. PLAAF and PLANAF success in previous interceptions suggests that the Chinese are being scrupulously careful to avoid giving the United States an opportunity to cry aggressor, apparently picking and choosing among the intruders and attacking only when there is a clear-cut opportunity for success at a minimum of risk. The last thing wanted by the air forces of Red China is an open fight with aircraft of the United States, for such a battle might well lead to all-out war. Only rarely do PLAAF aircraft find themselves completely caught out by American aircraft as was the Red Chinese MiG-17 near the Vietnamese border in May, 1966, and the Communist claim that the battle took place over Chinese soil was later all but confirmed by the United States.

Painful recognition of Red China's ability to damage American prestige by protests against claimed acts of war has led to a change in American policy and the open admission of accidental aerial intrusions by U.S. aircraft. In the summer of 1966, the United States admitted that the May battle might well have taken place over Communist China, but added that the intrusion had not been intentional. Later, in September, Washington stated that American aircraft had accidentally overflown southern China on two occasions while on missions over Viet-Nam during the month. On February 9, 1967, the U.S. Defense Department announced that an unarmed American naval aircraft had accidentally flown over the coast of Hainan earlier in the day, in spite of the fact that it was not known whether the aircraft had been tracked or whether any attempt at interception had been made. The new American policy of open admission, coupled with the border proximity restrictions imposed on American pilots in normal raid operations over Viet-Nam, has blunted the edge of the Red Chinese propaganda sword. These new policies are intended to officially assure the Red Chinese that the United States has no intention of willfully violating the air space of the People's Republic of China. Washington now recognizes that the Chinese are both capable and eager, under proper conditions, to defend their own skies and attack U.S. aircraft that get caught in their patrol net. The implications of accidental straying over Chinese territory assumed even greater significance with the initiation of raids over North Viet-Nam within a few miles of the Communist Chinese border in August, 1967. While such raids did occur, they were conducted only under Presidential order and under strict supervision of the

Administration in Washington and did not constitute relaxation of the standing twenty-five-mile rule.

Even under the conditions existing with border raids, the United States made every effort to avoid possible Red Chinese interpretation of such attacks as assaults against the People's Republic. When two American Grumman A6A attack planes from the carrier *Constellation* were reported missing after being forced by North Vietnamese MiG's and missiles on August 21, 1967, to fly over the Red Chinese border, the U.S. Defense Department immediately announced that the planes were presumed lost over the Chinese mainland. Five hours later, these fears were confirmed by a Chinese report that both aircraft had been shot down by the PLAAF over Kwangsi Province, but by that time the Chinese claim that the intrusion was an act of war could not be taken seriously. The American policy shift has repeatedly paid such dividends in the propaganda war, but it has not relieved border tension. Still not acknowledged in this "announcement war" is the U.S. unarmed high-altitude reconnaissance overflight program over the Chinese mainland, involving the communications monitoring "Flying Platform" flights of RB-47's in the 1950's and the atomic-oriented flights of the present decade. The current flights are whispered about in American quarters and grudgingly tolerated by the Red Chinese—presumably only until the day they are capable of bringing them to a halt.

The growth of Red China's defensive air force stems from the measures taken to counter the weakness of the PLAAF following the Korean War, the aerial defeats suffered at the hands of the Nationalists over the Quemoy straits in 1958, and the continual overflight and bombing threats of the early 1960's. The defense force was started with Russian help but soon became uniquely Chinese as the People's Republic became more isolated from the outside world. One of the last Russian acts before the final Sino-Soviet split was to provide the Chinese with a number of Mach-1.8 Mikoyan MiG-19 interceptors as well as with the capability to produce the new fighters at the National Aircraft Factory at Shenyang in Manchuria. The first Russian-built MiG-19's arrived in China in 1959, and plans to produce the fighters there were initiated. Although the MiG-19 was a difficult aircraft to fly because of its greater weight and twin-jet power plants, the original goal was to phase out the Shenyang MiG-17's and convert to full MiG-19 production as rapidly as possible. Plans for production of a Chinese version of the MiG-19 went back to October, 1957, when an agreement was concluded between the People's Republic of China and the Soviet Union for the advancement of Chinese Communist technology for the purposes of national defense. Then, on January 8, 1958, additional license agreements were signed in Moscow leading to increased activity at the National Aircraft Factory in 1958 and 1959. According to the Russians, the technical assistance given to the Chinese Communists in these two years alone exceeded that supplied during the entire first Five-Year Plan ending in 1957. The production objective this time was to do without Russian components and make the Shenyang MiG-19 truly an indigenous Chinese aircraft, with all elements of the aircraft, including the stressed supersonic airframes, jet engines, and necessary electronic gear, of Communist Chinese origin. The program had barely been started when the ideological problems of practical Communist politics brought it to a temporary halt.

The full scope of the Sino-Soviet split in 1960 was not realized by the rest of the world for at least two years, yet the disagreement was dramatic and conclusive. The Chinese regarded themselves as hard-line Stalinists and Mao Tse-tung as the world's leading Communist luminary. Nikita Khrushchev's denunciation of Stalin and the growing Russian policy of "peaceful coexistence" brought violent reactions from the Chinese. The Soviet answer in late 1959 was to cut off military aid and supplies to Communist China and present a bill for the equipment provided since the beginning of the Korean War. Chinese intransigence led to a complete break, and in August, 1960, the Russians pulled all their technicians and advisers out of China. According to Red Chinese sources, the Soviets withdrew 1,390 experts, canceled 343 contracts, and left 257 scientific and technical projects high and dry. Russian management and technical personnel at the National Aircraft Factory at Shenyang stripped their offices and even took their blueprints with them, leaving the new Shenyang MiG-19 production lines unfinished and inoperative. The Soviet pull-out coincided with a severe drought in Red China, so that the Chinese found themselves alone at a time of national crisis. In the face of the Russian move, the Chinese proclaimed that they could build their own MiG fighters and proceeded to do so. In late 1961, the first Shenyang MiG-19 fighters were completed; by the middle of 1962, a few had reached PLAAF service, with deliveries slowly increasing as the production problems were ironed out. By the summer of 1964, more than 100 were in PLAAF service, and by February, 1965, the numbers were climbing rapidly.

It was thought at first that the Chinese MiG-19 was somewhat lighter and did not have the sophisticated fire-control systems of the Soviet version. Later reports indicate that it is comparable in all respects. It still did not match the performance of the Mach-2 fighters then entering service in the United States and the Soviet Union, both now regarded by Red China as natural enemies. It quickly became apparent to the Chinese that improvements would have to be made or new models produced. To meet this new requirement, they went shopping—with mixed results.

In 1962, Red Chinese representatives of the second Ministry of Machine Building responsible for licensed aircraft production visited aircraft plants in Sweden and Switzerland in order to make arrangements for Chinese production of advanced aircraft designs under development in both European countries. Top priority was assigned to obtaining rights and assistance to produce an export version of the Swedish SAAB J-35B Draken, a modern Rolls-Royce RB-146-powered double-delta fighter-interceptor equipped with afterburners and a highly sophisticated electronic fire-control system. Negotiations fell through before any positive steps could be taken. In Switzerland, the Chinese showed interest in the AFA P-16 Mark III Strike Fighter, but nothing came of the connection. The Communists did, however, receive help in the area of machine tools and instrumentation, and National Trading Corporation purchases of precision instruments were made in France, Great Britain, Czechoslovakia, and East Germany. In January, 1964, navigation equipment was purchased in Japan, and the indigenous Red Chinese electronics and communications industry was developing its own avionics and electronic aviation gear.

Recognition of the People's Republic by France in January, 1964, seemed to offer another avenue of aircraft acquisition, and discussions at the French trade fair at Peking in April, 1964, centered on the sale of Dassault Mirage fighters and nuclear bombers to the Chinese. The possibility of a rapid PLAAF build-up with French aircraft troubled the Nationalists; yet, in spite of the need to replace the already obsolescent Shenyang MiG-19 fighters, no orders resulted. The reasons are unknown, but the Chinese Communists seemed to have an alternative up their sleeves. They also carefully investigated the possibility of Shenyang production of the French SUD Alouette III helicopter early in 1964, but once again apparently no sales were made. This plan may not be completely forgotten; a confirmed Chinese order for French-built Alouette III's was announced in Paris early in 1967.

The Communist Chinese aircraft industry in the early 1960's was much like that of Japan in the early 1930's when the Japanese began to convert to the production of aircraft of their own design. Communist China had joined the aircraft "manufacturing club" when the National Aircraft Factory was opened in the 1950's and when China, like Japan over thirty years earlier, originally produced aircraft under foreign licenses. The next development was to be the production of indigenous aircraft, and while it was still only a future hope for the Chinese, it was an absolute necessity if China was to maintain and expand the power of her air forces. The first order of business was to establish a production capability to meet the immediate needs of the PLAAF. To set the pace for this future development, the Academy of Military Sciences had been established in Peking in March, 1958, to direct defense research and military weapons development, including the design and development of aircraft. A number of new institutes under the Chinese Academy of Sciences were established in 1959, just before the Russian pull-out, to direct the development of the supportive equipment needed by the growing aerospace industries. The Institute of Automation and Remote Control and the Institute of Mechanics and Electronics, both in Peking, were directly related to the aircraft industry, while the Institute of Upper Atmospheric Physics at Wuhan was devoted solely to rocketry. These new institutes joined the Institute of Mechanics, which, since early 1956, had been under the capable direction of Dr. Chien Hsueh-shen, professor of aeronautics at M.I.T. between 1947 and 1949 and one of the world's leading aeronautical and jet propulsion experts, and which was responsible for rocketry and jet propulsion for Communist China's armed forces.

The growing political isolation of Communist China did not mean that the Chinese were completely cut off from Western technology. Much the reverse was true. Once their technical ties with Russia had been severed, the Chinese embarked on their own aeronautical research and development programs. Between 1953 and 1963, over 5,000 graduate aeronautical engineers were turned out by the Peking Aeronautical Engineering College, Communist China's top theoretical and practical school of aeronautical engineering, as well as Harbin Polytechnic University, Northwestern Technological University, Nanking Aeronautical Engineering College, and the other aeronautical institutions of advanced learning on the mainland. Aeronautical design teams were established by the Academy of Military Sciences in Peking, and two high-speed wind tunnels purchased in East Germany in 1959 were set up. By 1963 the growing profession of aeronautical engineering led to formation of the China Aeronautical Engineering Society under the direction of Shen Yuan, chairman of the board of directors of the Society and vice-president of the Peking Aeronautical Engineering College, and under the vice-chairmanship of Kuo Yung-hua, a deputy director of the Institute of Mechanics. Foreign aircraft were studied, including a North American F-86F Sabre flown to Red China on June 2, 1963, by its defecting Nationalist pilot; Lockheed F-104 Starfighters and Martin B-57 bombers of the Pakistan Air Force, and a De Havilland–Canada DHC-2 Beaver captured during the border fighting with India in 1962 and later returned to India as a gesture of good will. Western technical and aeronautical journals were also carefully studied, numerous Red Chinese technical journals covering jet propulsion, aeronautics, and avionics were published, and an aggressive effort was made to upgrade Chinese aeronautical technology. Dr. Chien Hsueh-shen, aeronautical editor for the journal of the Chinese Academy of Sciences, wrote that an understanding of foreign developments in aircraft and jet power was of critical importance to Communist China's future in the air. Dr. Chien further advocated mastery of German, French, English, Russian, and Japanese by Communist China's aeronautical engineers to enable them to keep up with international technical developments. Leading foreign texts in the field, as well as original Communist Chinese texts in aeronautical engineering, have been printed since 1961.

Four more ministries of machine building—the fourth, fifth,

sixth, and seventh—were established in 1963 and 1964 to handle the rapidly growing indigenous defense industries and produce the new equipment and aircraft originating under the auspices of the Academy of Military Sciences. The fourth, under Minister Wang Cheng, was responsible for communications and electronic equipment. Of particular importance to the PLAAF was the establishment of the seventh Ministry of Machine Building under Minister Wang Pin-cheng, then deputy commander of the PLAAF, whose new responsibility was to take over existing jet aircraft production and stimulate the development and production of original Red Chinese military aircraft to meet future defense needs. Each of the new ministers was appointed to the National Defense Council in Peking. By the summer of 1965 this was to pay handsome dividends, when China became an independent aircraft-producing nation.

While the aircraft industry struggled to overcome the loss of Soviet aid and direction, the PLAAF sunk to an all-time low in line strength and operational capability. Cancellation of Russian assistance and equipment was immediately felt in the PLAAF; new aircraft deliveries ended, and the small inventory of spares was soon used up. By the spring of 1964, combat-ready planes had dropped to 54 per cent of the total available aircraft. Even during the trying days of the Korean War, the maintenance rate had been held at 70 per cent. Aircraft were now scrapped for their parts to keep others maintained, and the Chinese desperately tried to make their own parts to keep their declining air force intact. The Chinese were also low on fuel (the oil fields in Manchuria were yet to be discovered). As a result, PLAAF training also suffered, and the flight time of its airmen dropped considerably, with a typical PLAAF fighter pilot logging 200 hours or less. Training and operational accident rates soared, further compounding the plight of the PLAAF. Because of the great dependence of the Chinese Air Force on Soviet parts and assistance, the

Russian move hurt the PLAAF more than any other Chinese military arm, and the competence of PLAAF pilots dropped appreciably lower than that of the Nationalist pilots on Taiwan or of the American pilots assigned to the Seventh Fleet, in evidence along the southern and eastern borders of Communist China. During the years between 1960 and 1964, Red Chinese aircraft inventory shrank, and morale in the PLAAF dropped at an alarming rate.

By the end of 1964, the paper strength of the PLAAF and the PLANAF consisted of 102 air regiments, assigned to 34 air divisions, and 6 independent air regiments. Approximately 1,770 out of a total of 2,600 aircraft were fighters, with only 120 of them less than four years old. About 340 of these fighters were nine years old or older, and most of the fighter forces were equipped with MiG-15's and MiG-17's in various degrees of combat readiness. The bomber regiments were equipped with about 300 IL-28 twin-jet bombers, while some 200 TU-2 attack bombers still remained in service. The force of TU-4 strategic bombers had dropped to about 20 aircraft. About 200 transports were in service, primarily IL-12, IL-14, IL-18, LI-2, and AN-2 types, as well as a few Tupolev TU-70 four-engine transport versions of the TU-4 for paratroop use. An additional 300 or so transports from the civil air fleet were available to expand the military airlift capability but, in the main, these aircraft were also obsolescent and of the short-haul variety. Some 180 trainers, consisting of YAK-11, YAK-18, MiG-15UTI, and IL-28U types, were assigned to training schools and line units. Even a number of obsolete LA-9, LA-11, PO-2, and YAK-12 types remained in service, and more than two-thirds of the PLAAF aircraft were over six years old.

Cut off from the outside world and with their aircraft-production plans facing enormous growing pains, the PLAAF and the PLANAF had to make do with what they had on hand. Fighter regiments were primarily equipped with Shenyang MiG-17 fighters (as late as the fall of 1966, many still were). Defensive units were stationed in a narrow belt along the coastline stretching from Russia in the north to the Gulf of Tonkin in the southeast, and bases were rebuilt and strengthened. This thin defensive line has left the interior underprotected, and all units are assigned secondary defensive areas in the event of enemy penetration of the outer defense belt.

Bomber regiments, still equipped chiefly with the obsolete Ilyushin IL-28, have concentrated on a maintenance level of training and on patrol duties but face the possibility of offensive operations at any time. Most PLAAF bomber units are stationed close to Red China's coastline, with the 8th Division facing Taiwan across the straits. While the PLAAF bomber force has never been used outside of mainland China, periodic maneuvers and night training flights are used to keep the force at combat readiness, although its lasting power against modern fighters would be very short indeed. It would appear that the primary function of the Red Chinese bomber force is to frighten the Nationalists on Taiwan and to convince the people on the mainland that the PLAAF is capable of striking out at the Nationalists or at American bases on the perimeter of Communist China at any time. The rumor circulating in Peking in June, 1966, that PLAAF bombers had attacked Taiwan fitted this mold, but no PLAAF aircraft has dared to fly over Taiwan since the Quemoy straits fighting of 1958.

More than 100 IL-28 bombers were also assigned to the PLANAF, the naval air arm; they were modified to carry torpedoes for the interception at sea of surface vessels threatening invasion of the coastal areas. These modifications were undertaken in the years when a Nationalist amphibious assault was considered a distinct possibility, and the aging IL-28's still constitute a long-range threat to any surface fleet, such as the U.S. Seventh Fleet, operating in the area.

The strategic bombers of the PLAAF have largely atrophied, and most of the few remaining TU-4's are used for long-range reconnaissance duties. Some effort, however, is still being made to maintain the force, specifically the 25th Air Division, located at Sian. This unit was selected for aerial drops of Red China's atomic bombs. Training for the aerial tests began in February, 1965, some three months before Red China's second atomic blast, on May 14, 1965. TU-4 bombers of the 25th Air Division practiced the dropping of dummy atomic bombs and executed evasion tactics to avoid being caught in the make-believe blasts. The third Red Chinese nuclear explosion, on May 9, 1966, of which color movies were shown in Japan early in 1967, revealed that this thermonuclear device was dropped from an aircraft, with the blast taking place at about 10,000 feet. The fact of Red China's hydrogen bomb, coupled with the dozen or so TU-4 bombers remaining to the PLAAF, hardly constitutes a modern nuclear-bombing capability, but it could be used against virtually defenseless neighbor states, a terrifying fact of life in the nations of the Far East.

Although the Chinese Communist hierarchy threateningly flaunts its nuclear capability on the world stage, the people of Red China seem to have a lackadaisical attitude toward their bomb. They are far more impressed with the achievements, real or implied, of their air forces. The PLAAF receives good press coverage in Red China and is in evidence everywhere as a helping arm. Its typical civil applications are the shipment of medicines and supplies to frontier regions and assistance during national emergencies. On the night of February 3, 1960, a PLAAF Ilyushin IL-14 transport flew in from Peking and dropped medicine urgently needed by a group of sixty-one food-poisoned road workers in Shansi Province within eight hours after a call for help had been relayed to the capital. All sixty-one lives were saved. In March, 1966, Shenyang MI-4 helicopters of the PLAAF were rushed to a commune earthquake disaster area to bring supplies and remove the seriously injured for hospital care. Other PLAAF activities, including participation in flying shows and sports events, have made the air arm the most admired branch of service in the PLA. The popularity of the air force–and the subsequent growth of an arrogant officer class outside of the Communist Party structure–was one of the determining factors in the supposedly morale-boosting move to eliminate distinctions of rank in the PLA and its branches. The order went into effect on June 1, 1965, and presently the officers and men in all Red Chinese military services wear the same uniform.

The removal of all insignia of rank and the loss of the distinctive PLAAF uniform have served only to further the decay in morale of the Red Chinese air forces. PLAAF morale problems have revealed themselves in several interesting ways, the most significant being the growing number of aircrew defections. While it is possible that many mainland Chinese would like to defect, either to Taiwan or just away from mainland China, few military men can do so. Air-force personnel, on the other hand, have a better chance, for they have aircraft with which to get away. The problems are many, however, for a pilot planning to defect must be able to endure "ideological training" and pass the loyalty examinations given in the PLAAF every three months in order to remain in the air force. Once this hurdle is passed, the prospective defector must consider the physical problems of getting away–particularly the wall of antidefection MiG fighter patrols along the coastline.

The first indication of the success of Nationalist broadcasts to the mainland offering sanctuary to defecting Communist pilots came on January 12, 1960, when a Russian-built MiG-15 fighter numbered 0651 was spotted at 8:26 P.M. by farmers of Ilan, in eastern Taiwan, after it had successfully crossed the straits. As the aircraft circled for a landing over the Nanao riverbed, the Taiwanese were startled to see the Communist red stars on the wings. The MiG then dropped its landing gear and came down along the riverbed. As it came in, the nosewheel struck a large rock and the MiG overturned and burst into flames. The pilot, dressed in civilian clothes, was killed. He carried no identification. The event revealed that escape was possible. The Nationalists announced the flight to the world and beamed the story of the defection to the mainland to inspire other PLAAF pilots to escape. To make the effort doubly worthwhile, the Nationalists repeated their offer of payments in gold for various PLAAF aircraft: 500 ounces, or U.S. $17,500, for transport aircraft such as the AN-2; 1,000 ounces, or U.S. $35,000, for a MiG fighter; and 4,000 ounces, or U.S. $140,000 for a bomber.

The loss of the MiG-15 prompted the PLAAF to circulate a confidential report on the flight to PLAAF unit officers along the coast early in 1960. The defection flight had originated from a PLAAF fighter regiment at Luchiao, in Chekiang Province, some 200 miles north of Taiwan, and the purpose of the document was to strengthen security measures within this and other units to avoid repetition of the event. The MiG counterpatrol was established, using pilots of unquestioned loyalty, and ideological training was increased. It seemed to work, for no defectors were reported for more than a year, although it is possible that other attempted escapes were thwarted. Then, on September 15, 1961, a civil Shenyang AN-2 left Kiaochow Air Base on a crop-dusting mission in Shantung Province, crossed the coastline at low level, avoiding radar discovery and the feared MiG patrol, and crash-landed in a rice paddy in South Korea. The pilots, Shao Hsi-yen and Kao Yu-tsung, requested and received asylum on Taiwan, and in December the AN-2 was shipped to the island for display. These were men who had grown up under Communism in China. Both had received their training in the PLAAF, and Shao had become a MiG-15 pilot when he left the Fifth Aviation School at Tsinan. The defection had been carefully planned and executed, and once again the Red Chinese were bombarded with Nationalist propaganda. The defecting pilots were heard over the Nationalist radio, and the story of their welcome on Taiwan was widely reported.

Reactions to the broadcasts of Shao and Kao were unknown until a dramatic defection on the afternoon of March 3, 1962, when Lieutenant Liu Cheng-sze, a twenty-five-year-old fighter pilot in the PLANAF, flew his naval MiG-15bis fighter directly to Taoyüan Air Base on Taiwan escorted by a protective umbrella of Nationalist F-86F fighters. Liu had also graduated from the Fifth Aviation School at Tsinan, receiving his wings in 1960, and first thought of defection when he read a report of the MiG flight from Luchiao in 1960 (the report had omitted the fact that the pilot had been killed upon landing). Liu had also heard the Nationalist radio broadcasts and, significantly, was stationed at Luchiao in the 16th Regiment of the 6th Air Division of the PLANAF, attached to the Eastern Sea Fleet. After months of planning, Liu made his break during a training flight, dashed for Taiwan, and was soon intercepted by Nationalist Air Force fighters. After rocking the wings of his MiG-15, in accordance with the defection procedures broadcast over the radio months before, Lieutenant Liu was directed to Taoyüan Air Base, where he was welcomed with open arms. The Nationalists now had their first Red Chinese fighter. Soon it was being test-flown in Nationalist insignia, and later it was placed on display.

The defection of Liu Cheng-sze had far-reaching effects on the PLAAF, and steps were taken to maintain a close watch over air-force personnel. The loyalty tests were initiated, and pilots were checked to be sure that they had the proper "class background." Any deviation from the established mode of thought led to grounding and, in some cases, dismissal from the PLAAF. Political officers were assigned to the PLAAF in greater numbers, and daily life in the air force came under close supervision. The Red Chinese also made counterdefection offers in July, 1962, promising to pay 700 ounces of gold, or U.S. $24,500, for a Nationalist C-47 transport, and up to

8,000 ounces of gold, or U.S. $280,000, for a Lockheed U-2 reconnaissance plane. The higher prices offered by the Communists did not seem to turn the tide of defections, although a single Nationalist pilot, Captain Hsu Ting-tse, flying an F-86F Sabre of the 2d Squadron of the 43d Regiment, did defect from Taiwan on June 2, 1963, and land safely on the mainland.

The centennial of the birth of Dr. Sun Yat-sen was perhaps the most important single celebration in modern Chinese history. Although Dr. Sun is recognized by both the Communists and the Nationalists as the father of the Chinese Revolution, the date of his centennial differed by a year in the two rival governments, with the Nationalists celebrating on November 12, 1965, and the mainland Chinese in 1966. Months of preparation on Taiwan were coming to a head on the day before the centennial celebration, and in the early afternoon of November 11, as President Chiang Kai-shek addressed a group of government officials at Taipei, the capital of Taiwan, he was approached by an excited aide, who then whispered in his ear. President Chiang announced to the group that the first Chinese Communist bomber had defected to Taiwan. Within minutes, the news was on the radio; within the hour, newspaper extras were on the street.

It had been more than three and a half years since the last successful defection to Taiwan, and now an Ilyushin IL-28 bomber from the 22d Regiment of the 8th Air Division was intact in Nationalist hands, flown not by a single escaping pilot, but by a crew of three men who had all desired to leave the Communist mainland. All three–twenty-eight-year-old pilot Li Hsien-pin, navigator Li Tsai-wang, and gunner Lien Pao-sheng–had successfully passed the PLAAF's battery of loyalty examinations and had joined together in their dangerous plan. It was a monumental propaganda coup for the Nationalists, and when Li Hsien-pin, the pilot, spoke at the Sun

Centennial Rally the next day and denounced Communism, the story gained world-wide attention.

The escape plane had started at noon from Hangchow as part of a group of 8th Division IL-28 bombers on a practice bombing and strafing mission. The defecting IL-28 remained close to the coast, and the moment it broke away at top speed for the dash to Taiwan, the Communist alarm was out. But it was too late. Within minutes, the bomber was spotted by radar on Taiwan and intercepted by eight Nationalist F-104 Starfighters. When pilot Li waggled the wings of his IL-28, the Nationalist pilots knew his intention and escorted the plane to Taoyüan Air Base, where it skidded to a crash landing in a meadow, wounding all three crew members. Lien Pao-sheng, the gunner and radio operator, died the next morning.

PLAAF headquarters at Peking were quickly alerted to the disaster, and Vice-Commander Cheng Chun flew to Hangchow the same evening to meet with officers of the 8th Division. All training and patrol flights were terminated at Hangchow until further notice. In addition, all flying activity at the other bases within close proximity of Taiwan–Luchiao and Chowshan in Chekiang; Amoy, Foochow, and Tungshan in Fukien; and Paiyun and Sanyia in Kwangtung–were severely curtailed. Suspicion replaced efficiency, and other defection attempts were made or imagined. At least one other PLAAF pilot was successful in an escape attempt. In this case, Lieutenant Liang Kuo-hsin of the 103d Regiment of the 24th Air Division defected to Macao and ultimately reached Taiwan by way of India in time to attend the Nationalist Freedom Day Rally, on January 23, 1966. Then, in March, 1966, former captain Cho Yung, a PLAAF mechanical engineer prior to his reassignment to the PLANAF at Canton in 1963, made it to Hong Kong with his wife in a power launch, finally reaching Taiwan in January, 1967. To this day, the Nationalists broadcast defection instructions to the mainland and attempt to persuade PLAAF pilots, crews, and ground personnel to fly to Taiwan for "happiness and a bright future." During the height of the reported unrest brought on by Mao Tse-tung's Cultural Revolution, the Nationalists even went so far as to appeal to complete units of the PLAAF to resist Communist rule, promising that they would receive formal designation as detachments of the Nationalist Air Force.

The combined problems of equipment obsolescence and sinking morale haven't kept Red China from making use of its air power in other areas of the world whenever opportunities for political and military advantage presented themselves. In the early 1960's, the Chinese Communists undertook the responsibility of re-equipping and re-creating the Korean War vintage air force of the People's Democratic Republic of Korea, on the theory that a strong air force in Korea would be an effective first-line shield of mainland China and an integral part of the Chinese Communist air-defense system. Red China supplied hundreds of aircraft to North Korea in spite of the restrictions against such aid in the Korean truce agreements. The North Korean Air Force was organized on the Chinese pattern, with the bulk of its strength allocated to fighter regiments. The Chinese supplied 380 MiG-15bis fighters and fighter-bombers and Shenyang MiG-17 interceptors, the efforts of a full year of production at the National Aircraft Factory. The planes were flown into North Korea to equip a fighter-bomber air division and six fighter air regiments, the latter formed into three air divisions of two regiments of about thirty-five aircraft each. Former PLAAF IL-28 bombers, as well as IL-28U trainers, were also shipped across the border to equip a light-bomber air division. It wasn't until the middle of the 1960's, when North Korea appealed to the Soviet Union for aid and turned against Red China, destroying Peking's first-line-of-defense theory and negating its massive aid program, that the silver cord of Chinese control over the air force of North Korea was broken. But the Chinese found other areas of opportunity and made good use of their aircraft to extend their sphere of influence.

Late in 1963, the Chinese Communists presented the government of Nepal with three Russian-built Lisunov LI-2 transports to provide military airlift capability to this bordering state. A Shenyang AN-2 Fong Shou No. 2 was also presented to the king of Nepal for his Royal Flight, and Red China was officially in the business of exporting Chinese-produced aircraft to non-Communist nations. When a second Chinese-built AN-2 was supplied to Nepal, the two Shenyang biplanes were taken over by the Royal Nepal Airlines and put into charter services, serving as testimonials to the durability of aircraft produced by the Chinese Communists.

In August, 1964, the Chinese supplied North Viet-Nam with about fifteen MiG-15bis and MiG-17 fighters, forming the basis of Hanoi's air force. Vietnamese pilots were trained in China, and PLAAF advisers were dispatched to Hanoi. In 1965, when Albania asked for aid for its air force, Chinese-produced parts and components were shipped to Tirana to keep that nation's four squadrons of Russian-built MiG-15bis and MiG-17 fighters maintained and in the air. Then, in 1966, thirty Shenyang MiG-17 fighters were sent to Albania to equip two interceptor squadrons and replace the older Russian fighters then in use, an indication that the problems of aircraft availability were slowly being solved on the Chinese mainland.

It was soon obvious that the Red Chinese were searching for export markets for their older aircraft, a sign that newer models were in the offing. Conclusive proof came with General Liu Ya-lou's tour of nations friendly to Red China. As commander in chief of the PLAAF (a post he held since the official formation of the Air Force in 1949), General Liu and his entourage toured the air-force facilities of Rumania, Albania, Pakistan, Cambodia, and North Viet-Nam and established a series of aircraft-aid agreements with each nation except Rumania early in 1965. The trip was brought to an untimely end when General Liu was caught in the first American air raid north of the 17th parallel in North Viet-Nam. When he died in Shanghai on May 7, 1965, as a result of the wounds received in the raid, the Chinese announced he had died of a disease, making no mention of the U.S. air attack. In September, 1965, Liu was replaced by Wu Fa-hsien, a political appointment that sidestepped Wang Pin-cheng, Liu's former deputy commander and then minister in charge of the seventh Ministry of Machine Building; and the troublesome PLAAF at last came under full Party power. Wu had been assigned to the PLAAF in the early 1950's as its Deputy Political Commissar, and became the full Political Commissar in 1959. Although Wu was untrained in the management of PLAAF operations and functions, his political connections gave him great powers, and the PLAAF became more than ever before an instrument of Party policies.

The air agreements set up by Liu before his death became key elements in Communist China's foreign policy, and the Red Chinese were soon playing the former Russian role of advisers and suppliers of aircraft. More MiG-17 fighters were sent to North Viet-Nam. By late 1965, Hanoi had more than sixty Red Chinese MiG-15bis and MiG-17 fighters on hand for use against attacking U.S. aircraft, and Soviet aircraft aid was just beginning to be felt. The relationship between the PLAAF and the air arm of North Viet-Nam became a very close one, for both were facing a common enemy in the United States. PLAAF installations in close proximity to the Vietnamese border became training grounds for the North Vietnamese air forces. Red Chinese bases, such as the PLAAF complex at Mengtsz in Yunnan Province, entered service as major maintenance and repair points for North Viet-Nam's MiG-15bis, MiG-17, and Russian-built MiG-21 fighters as well as for its Russian-made IL-28 bombers.

This close partnership of interest has led to a growing American concern over the possibility of North Vietnamese fighters' operating out of Red China as a result of the bombing of their home bases in and around Hanoi and Haiphong, although such use would put North Vietnamese fighters at a great disadvantage because of the ranges involved and the complications that the PLAAF would face if an American policy of "hot pursuit" was authorized. Within days after the first American attacks on North Vietnamese fighter fields near Hanoi, in late April, 1967, the Chinese Communists claimed that PLAAF fighters had downed two USAF Phantoms over Kwangsi Province and had suffered no losses. Since then, many additional claims have been made about the shooting down of U.S. aircraft over China, many of them ignored or denied by American sources. The fear of hotter U.S.–Red Chinese involvement as a result of the operation of American aircraft over Southeast Asia continues to be a very real one.

Red Chinese air aid to Pakistan began in July, 1965, with the expansion of existing Pakistani Air Force facilities and the construction of new airfields in East Pakistan under PLAAF supervision. By September, 1965, Red Chinese IL-14 transports were flying almost daily between Canton and Karachi with arms, advisers, and equipment. This aid program was temporarily forestalled by the war between India and Pakistan in the summer of 1965. When the smoke cleared, the Pakistan Air Force had lost one-third of its aircraft. The Chinese were on hand to rebuild the force, while Great Britain and the United States refused to sell more aircraft to Pakistan. Within a few months, Pakistani pilots were being sent to Red China for training and deliveries of Chinese aircraft began. In March, 1966, when the results were first revealed to the outside world during a military display in Rawalpindi on Pakistan Day, foreign observers were shocked. They had expected to see Chinese aircraft, perhaps Shenyang MiG-17 fighters. But when the Pakistan Air Force flew overhead, it was led by four highly polished Shenyang MiG-19 fighters in Pakistan insignia, followed by the American-built F-104 Starfighters, F-86 Sabres, and B-57 bombers of the Pakistan air arm.

Later in the year, following a $30 million–$40 million military aid agreement whereby China would supply Pakistan with 125 Shenyang MiG-19 and MiG-17 fighters, Chinese deliveries were increased to the point where the Pakistan Air Force was able to establish four new fighter squadrons, flying eighty Shenyang MiG-19's, and back up its bomber units with a new eight-bomber squadron equipped with eight Chinese-supplied Ilyushin IL-28 bombers originally built in the Soviet Union, with two additional IL-28 aircraft in reserve. Chinese aircraft availabilities and production at the National Aircraft Factory had obviously come a long way, and Red China was now able to export some of its most modern aircraft without critical loss to the PLAAF.

Red Chinese aid to Cambodia was more modest. Following the signing of a military-aid agreement in October, 1965, PLAAF advisers reorganized Royal Khmer Aviation, Cambodia's military air arm. Cambodian pilots were sent to the People's Republic of China for conversion training in Chinese MiG-15UTI advanced jet trainers, and in February, 1966, eight Chinese MiG-15bis fighters of Korean War days, enough to set up a Cambodian fighter-bomber unit, were flown into Cambodia from Kunming in Yunnan Province. This was soon followed up by the delivery of a dozen Shenyang MiG-17 fighters, giving American observers in Cambodia their first opportunity to take a close look at the products of the National Aircraft Factory. One reported that the Chinese fighters appeared to be based on the early Russian fighter version of the MiG-17, while another noted that they lacked the sophisticated electronic gear of the later models of the MiG-17 produced in the Soviet Union. These new MiG's were used by the Cambodians to equip an interceptor squadron. Both Cambodian squadrons are retained at Pochentong

Airfield in Pnompenh, and the aircraft are painted light gray overall in typical Chinese style. The PLAAF maintains a permanent mission attached to Royal Khmer Aviation that is also stationed at Pnompenh, right across the field from the American area.

The extent of the Red Chinese aid to Cambodia at one time prompted Prince Sihanouk, the Cambodian chief of state, to say that Peking was his country's "number one friend." The Soviet Union also qualifies as a "friend," for in 1967 the Russians supplied Cambodia with five MiG-17 fighters to augment the Red Chinese equipment. The avowed Soviet purpose was to strengthen Royal Khmer Aviation, although it would appear that the international competition between the Soviet Union and Communist China for spheres of influence led to the delivery, rather than Chinese lack of ability to supply more aircraft.

In spite of the improvement in relations between Communist China and neighboring states, the Chinese are still faced with the frustrating reality of free-ranging American and Russian aircraft along their borders. To combat this menace, the Chinese have enlarged existing facilities and built new air bases in Manchuria, facing Russia and Mongolia, and in a semicircle that stretches from the Luichow Peninsula and the island of Hainan through southern China to the massive bases at Mengtsz, forty miles from the North Vietnamese border, and Nanning, eighty miles from the border. Since May, 1965, air exercises have been held frequently along the southern Chinese defensive belt and at the three main airfields on Hainan. The entire Canton Military Region is on ready alert, with upwards of 200 fighters ready for interception use at any time. Some airfields, such as Fucheng, can now handle fifty or more jet fighters, and maintenance and repair points, such as Linshui Field, keep the force well supplied and serviced. These fields are well protected by ground-to-air missiles and manned interceptors and constitute a major threat to any potential air attack on the mainland from the south.

Pilotless American Ryan BQM-34A Firebee reconnaissance drones, launched toward Red China from Lockheed DC-130A Hercules mother planes over South Viet-Nam, are having an increasingly difficult time penetrating these fighter defenses. Designed to give the USAF a low-level view of the defense positions in southern China, the drones have become a prime target for the PLAAF regiments in Yunnan, Kwangsi, and Kwantung provinces, and the Red Chinese claim to have downed nine of them between November 15, 1964, and the spring of 1966, three of which are on display in Peking. It would appear that the unofficial air struggle between the United States and Communist China is virtually stalemated, and nothing short of a dramatic shift in the power balance or direct Red Chinese involvement in the Viet-Nam fighting will alter it.

Any significant increase in the defensive and offensive power of the PLAAF in southern China presupposes that the dual problems of aircraft obsolescence and pilot inefficiency have been overcome. There are signs that this is exactly what is happening in Communist China and that the strength of the PLAAF is once again improving. While the southern PLAAF fighter regiments are still largely equipped with Shenyang MiG-17 and MiG-19 fighters, a growing number of modern delta-winged fighters have been spotted on the Chinese fields and in the air. First seen in February, 1965, the new fighters indicate that the PLAAF is in the throes of a major re-equipment program, one that will bring the Communist Chinese air arm more in line with the modern air forces it opposes.

The story of the PLAAF's new fighter goes back to 1960, when the Soviets terminated their aid to Red China. While they were delivering Mikoyan MiG-19 fighters to Communist China, the Russians were also making arrangements to export their new Mikoyan MiG-21 fighter to India, Red China and other Communist states, and Finland, as well as to aid India and Red China in the license production of the MiG-21. A few MiG-21 prototypes were delivered to China before the final Sino-Soviet break, but by the time of the schism, nothing had been done to aid the Red Chinese in its production. With the Chinese already facing apparently insurmountable production problems with the Shenyang MiG-19, it seemed unlikely, if not impossible, that the MiG-21 samples would ever be of any use to them. With the announcement in 1962 that the MiG-21 would be produced by Hindustan Aircraft in India and that Russia would not supply Red China with these aircraft in quantity, the scope of the Sino-Soviet split became obvious, and the termination of Russian aid to China was soon common knowledge. The Chinese, faced with the future use of the MiG-21 by India and an obvious disparity of advanced aircraft, vowed to produce their own MiG-21 fighters. They soon found themselves unable to do so, for they were lacking in knowledge and facilities.

Chinese production of the MiG-21 fighter remained a forlorn hope, and while the Chinese shopped around the world for another fighter and became more proficient in modern aircraft technology, the few Soviet MiG-21 prototypes gathered dust in China. As the months passed, Chinese engineers became experts in the technique of taking apart, rebuilding, and even improving sophisticated foreign machinery and equipment. At what point the Academy of Military Sciences design and engineering staff decided to tackle the MiG-21 is unknown, although Chinese sources say that it took five years to re-create the aircraft along Chinese lines and put it in production. The important fact is that they did produce an aircraft that is equivalent to the late-model MiG-21*f* fighter, one equipped with heat-seeking air-to-air missiles. The aeronautical and jet engineers responsible have not been identified, and no public announcements have been made regarding the presentation of awards for "inventions, technical improvements, and rational suggestions pertaining to production" so rewarding to designers in the People's Republic of China. But there is little doubt that an elite group of Communist Chinese engineers will be presented with the appropriate medals and cash awards as a result of their work on the Academy of Military Sciences adaptations of the MiG-21.

The first announcement of the unqualified success of the "Chinese-copy" MiG-21 was made at a secret session of the National Defense Council in Peking on January 8, 1965. General Wang Pin-cheng, who had been second in command of the PLAAF under General Liu Ya-lou and was now newly appointed minister of the seventh Ministry of Machine Building responsible for indigenous Red Chinese military aircraft, reported on the production plans for the new PLAAF fighter, a new jet bomber, and a series of land-based short-range nuclear missiles. Minister Wang had been entrusted with the task of expanding aircraft production as part of the newly initiated "five-year program of self-reliance" originally scheduled to start in 1963 but repeatedly postponed because of domestic economic and technical problems. Communist China's productive capability had finally started to come of age, and the fruits of its labors would soon be at hand. By March, 1965, about twelve to fifteen Red Chinese MiG-21 fighters had entered evaluation service. Within the next year and a half, more than 100 of the 1,400-mph fighters had been assigned to PLAAF fighter regiments.

The production of MiG-21 fighters in Communist China became one of the biggest news stories on the mainland, and the Chinese man in the street is aggressively proud of the PRC's ability to produce a modern fighter in the McDonnell F-4C Phantom class without direct aid from other nations. Shenyang MiG-21 fighters have been spotted in increasing numbers over Hainan and southern China. In January, 1966, during the American bombing lull over North Viet-Nam, companies of PLAAF Shenyang MiG-21 fighters were rotated between six airfields in Red China and the airfields north of Hanoi in North Viet-Nam. Obviously intended as a show of Communist strength for the Vietnamese, these aircraft cleared out of North Viet-Nam as soon as the American raids were resumed at the end of the month.

In spite of the anger felt by the Red Chinese over the operation of U.S. aircraft so close to their borders, the air war over Viet-Nam turned out to be a technological boon for China. In a case of clever diplomatic duplicity, it provided the PLAAF and the aeronautical design staff of the Academy of Military Sciences with an up-to-the-minute parity check and quality-control bench mark of the Shenyang MiG-21 design and its production. Since the early 1960's, when the PLAAF first provided the North Vietnamese with aircraft and training, the two Communist air arms had had a close relationship at the command level. Even when deliveries of aircraft began to reach Hanoi from the Soviet Union, the Chinese maintained their position of influence at the top. Red China, in fact, complained that the Soviets were niggardly in their aid and insulting in their offerings, claiming that the Russian aircraft were old models and badly maintained.

The Soviet position was difficult, for while shipments to Hanoi via surface vessels were allowed through the American picket lines around the North Vietnamese seaports, Russian deliveries of major offensive and defensive weapons might be stopped at any time. As a result, the Soviet Union was forced to turn its military-equipment shipments for North Viet-Nam over to the Red Chinese for transfer to Hanoi through the Chinese mainland by means of trucks and small-gauge railroads. As the fighting in Viet-Nam continued through 1966, the Russians began shipping late-model MiG-21*f* day fighters and MiG-21P*f* all-weather interceptors to North Viet-Nam via the Chinese supply line, with the shipping crates clearly marked. Russian checkers in the Democratic Republic of Viet-Nam soon discovered discrepancies in the shipments as they arrived, for many of the crates were re-marked in Chinese characters, suggesting that the military aid had come from Red China. More pointedly, some of the new Russian fighters were missing, and in their place were substitutions of Red Chinese aircraft. During this period, a number of MiG-19 fighter types were reported to have shown up on North Vietnamese airfields and may have originated in China. Such aircraft were already being exported to Pakistan by the Chinese Communists.

It was not until February, 1967, that the Russians openly complained about the Chinese switching. When they finally did, by means of a press announcement in Moscow, they accused the Chinese of stealing supersonic MiG-21 fighters intended for Hanoi and replacing them with "obsolete and well-worn" PLAAF fighters. By way of confirmation, Nationalist Chinese sources reported in March that Red China had recently supplied the North Vietnamese with sixty obsolete MiG-15bis and MiG-17 fighters. The Russians further accused the Chinese of betraying the interests of the Vietnamese by substituting maintenance equipment and supplies, of long delivery delays en route, and of "forgetting to replace" ground-to-air missiles and other military equipment removed from the shipments. The problem was not resolved until later in the spring, when an agreement was finally thrashed out between the Soviet Union and the People's Republic of China whereby North Vietnamese guards would assume physical control of the Russian shipments at the Chinese border to ensure safe delivery of the war goods through Red China. It is clear that the Chinese sorely abused their relationship with the Democratic Republic of Viet-Nam when the Chinese national interest was at stake. It is not known how many Russian fighters were siphoned off of the supply line, but the actual numbers are unimportant. What is important is that the Russian MiG-21*f* and MiG-21P*f* fighter models will provide the Academy of Military Sciences and ultimately the seventh Ministry of

Machine Building with a raft of new design ideas and improvements for the Shenyang MiG-21. The technical and design transfusions may well lead to the development of further new Red Chinese interceptor aircraft.

Of equal importance in the Communist Chinese five-year program of self-reliance is the development and production of a new PLAAF jet bomber to replace the IL-28, fewer than 160 of which remained in service in China by the summer of 1967. Yet little has been heard of this aircraft. The delivery of a few Tupolev TU-16 swept-wing jet bombers to Communist China in the early 1960's—as to Indonesia—has been reported, but Russian sources refuse to confirm this. The development of a National Aircraft Factory version of the TU-16, or a completely new bomber of Red Chinese design with similar or superior characteristics, would give the Chinese bomber regiments an aircraft capable of striking deep into Russia or reaching far afield of the mainland, bringing all of Japan and the U.S. bases on the Asiatic mainland and in the Pacific within bombing range. Late in 1964, the Chinese claimed that they already had a means of delivering their atomic bomb, and any improvement in this capability could dramatically alter the balance of power in the Far East.

The new fighter and bomber programs came just at the time that the PLAAF faced dissension within its ranks. Some air-force officers were, according to Chinese Communist sources, "losing their revolutionary spirit" by claiming that the weaponry of the PLAAF was no match for superior American and Russian technology. These so-called revisionist elements were opposed by the hard-line Communists, who regard "the thought of Mao Tse-tung" as the solution to all their problems. PLAAF companies found themselves split between pilots who carried copies of the small red book *Quotations from Chairman Mao Tse-tung* with them at all times and those who felt that changes should be made along Russian lines. This dichotomy of Red Chinese opinion reached throughout the

People's Liberation Army and the entire national political structure. The growing threat of revisionism led to a dramatic attempt by Mao Tse-tung to rekindle the flame of revolution in every Chinese heart and mind. Facing the severest political test of his life, Mao personally led the entire nation into an orgy of revolutionary recommitment, which he called the Cultural Revolution.

The Cultural Revolution appears to have been started in Shanghai in November, 1965. By the fall of 1966, there were reports of growing internal opposition to the new movement and of a planned coup to overthrow Mao. By December, news of civil unrest and armed clashes burst out of Red China to stun the world. There were even speculations that Communist China might be headed for civil war. However, there were no indications that the PLAAF was seeing combat duty or that the air forces of Red China were being used in any special way. It was almost as if the air arm were waiting on the side lines to see how the ideological conflict came out. Without question, the PLAAF felt the strain and the effects of the Cultural Revolution. Early in January, 1967, it was reported that Red Guards, members of a militant youth group that rampaged all over the mainland at Mao's urging, attacked workers at the National Aircraft Factory at Shenyang. This might well have delayed Shenyang MiG-21 production and delivery.

PLAAF regiments were alerted to be ready for action against any organized domestic opposition. They were also put on guard against any Nationalist or Russian attempt to take advantage of the situation in China. Extra PLAAF fighter patrols along the 4,000-mile Sino-Soviet frontier, the longest common border in the world, kept Chinese fighter forces in the north tied up for months. This activity is likely still continuing, since further mobilization orders were issued directly by Mao Tse-tung early in 1967. On August 6, 1967, the Chinese further aggravated the situation by nullifying long-standing agreements with both the Soviet Union and North Korea that provided for mutual assistance to aircraft in distress, or for those that might have inadvertently crossed the now touchy borders.

These extra patrols led to an engagement with the Nationalist Air Force fighters over the Quemoy straits, the first air combat in this area since 1961. The air battle started a few minutes after 1 P.M. on January 13, 1967, when twelve Red Chinese Shenyang MiG-19 fighters took on four American-built F-104 Starfighters from Taiwan on a routine patrol northeast of Quemoy. The assault was a complete surprise. Although the Nationalist fighters had spotted the MiG's some distance away and were ready for combat, since 1961 such sightings had always resulted in a Communist flight from the area. But not this time. In three groups of four aircraft each, the Red Chinese fighters closed in for the attack, cannons firing as they came. The fight was over in less than two minutes, and the Nationalists claimed that the Communist jets disengaged when two of their number were quickly downed by the missile-equipped Starfighters. It was the first time that Nationalist aircraft had met Red Chinese fighters of the MiG-19 type in the air.

This attack by Chinese Communist aircraft had obviously been ordered. The incident had the earmarks of a planned Red Chinese attempt to create an outside military diversion to take people's minds off their troubles at home. If the PLAAF defeat was as rapid and conclusive as it seems to have been, the only possible result would have been an additional lowering of PLAAF morale. Although the engagement galvanized the Nationalists into renewed watchfulness over the straits, the Red Chinese performance was not soon repeated. The point was made, however, that action over the straits can and will take place only when the PLAAF and the PLANAF are instructed to attack or to hold their ground.

Other, less militant appearances were made by the PLAAF

as the Cultural Revolution cooled down somewhat early in 1967, only to explode again in later months. In February, 1967, a flight of PLAAF aircraft showered a mass rally in Peking with leaflets lauding Chairman Mao Tse-tung, and the air force participated in a similar demonstration at Foochow in Fukien Province, along with units from the PLA and the navy. As fighting broke out in the summer of 1967, the PLAAF was finally brought into action on internal missions. Late in July, air force transports dropped paratroops at Wuhan to quell anti-Mao demonstrations, and pro-Mao PLAAF units were placed on alert for possible use in the future. In spite of these apparent commitments to Mao's philosophy, underground opposition to it remains a factor in the air forces of Red China and may reveal itself unexpectedly from time to time. Late in May, 1967, Defense Minister Lin Piao, a staunch Mao supporter dedicated to the Cultural Revolution, initiated a purge of anti-Mao personnel in the armed forces of the People's Republic of China, claiming that "malicious elements" had all but taken control of the PLAAF and PLANAF. By the end of the month, Lin Piao's crusade had reached the very top command of China's air arm with the removal of Chang Ting-fa, then deputy commander of the PLAAF, from his post. In a matter of months, the only PLAAF staff officers that held their posts were the commander in chief, one deputy commander, and the first political commissar. However, this drastic action does not seem to have allayed the unrest, for, late in the summer of 1967, a number of air force units still refused to support Chairman Mao's political purges. Instead, they boldly reaffirmed allegiance to Liu Shao-chi, President of the PRC and all but discredited by Chairman Mao as the personification of the "reactionary" opposition to the Cultural Revolution.

Even the Civil Aviation Administration of China came in for its share of attention during the unrest. CAAC offices in Peking were taken over by PLA forces early in February,

1967. The army also occupied CAAC offices throughout the country as well as airfields and flight-training facilities. These seizures reportedly were made to ensure normal flight operations on domestic and foreign CAAC routes and to aid the progress of the Cultural Revolution, although an announcement from military headquarters stated that the move was also a precaution against the eventuality of war. In spite of these precautions, by July, 1967, air travel on the mainland was reportedly nearing a complete breakdown because of interference by anti-Maoists. Even Premier Chou En-lai's aircraft was rerouted to a guarded airfield when he flew to Wuhan in July to plead for the release of two government officials being held by a PLA unit loyal to President Liu Shao-chi. This protective step was taken following the discovery of a plot by anti-Maoists to kidnap Premier Chou upon his landing at the Wuhan airport.

As the Cultural Revolution reached out across the mainland, passengers on CAAC airliners were thoroughly exposed to the thought of Mao Tse-tung. Russians returning home at the end of 1966 stated that CAAC cabins were often decorated with the inevitable portrait of Chairman Mao, while the walls were lined with selections from his writings. CAAC hostesses hurried through their duties to allow sufficient time to read and then sing more Mao quotations. In some cases, this performance was followed by a radio broadcast that further praised Chairman Mao and the Cultural Revolution. This overwhelming personality cultism is reaching into every level of Red Chinese life. Success in every venture is being credited to the application of Mao's philosophy. Even the Nationalist Lockheed U-2 reconnaissance plane retained on permanent display at the Chinese People's Revolutionary Military Museum in Peking is now said to have been downed by Mao's thought, whereas before the days of the Cultural Revolution it was reported that the U-2 had been hit by anti-aircraft fire. But the new approach is the official one, and even the museum guards will not now concede that a weapon of any kind had a bearing on the plane's destruction.

It is hard to believe that a modern air force can continue to operate efficiently or long when such a rigid philosophy is imposed on its personnel. Certainly the failures of the PLAAF and the PLANAF will be hidden, and the successes will be snatched away from the men who achieve them and the credit given to Chairman Mao. With the devaluation of the Chinese air forces, their skill and aircraft becoming mere adjuncts to Maoism, the future development of Red China in the air remains in question. Yet there is little doubt that the Chinese can produce modern aircraft in quantity. Designers and engineers on the mainland are presently working under the direction of the Military Science Academy on new and original models, intended for production at the National Aircraft Factory. As soon as the designs are approved, they come under the jurisdiction of the seventh Ministry of Machine Building—the governmental agency responsible for expediting construction problems and delivering the aircraft to the PLAAF. The very nature of the Chinese approach to aircraft design and development virtually guarantees the inferiority of the new Chinese Communist aircraft to contemporary American or Russian models. The almost mystical belief in the ability of the Chinese "workers" to invent and produce the advanced aircraft needed by the People's Republic of China must certainly suffer severely when faced with the burgeoning technical advances of the United States, as evidenced by the Lockheed SR-71 now used over the Chinese mainland on reconnaissance missions, and the array of modern aircraft first displayed by the Soviet Union at the now famous Domodedovo Air Show in the summer of 1967. The closing of aeronautical engineering schools and other institutions of higher learning all over the mainland as part of the Cultural Revolution could only lead to further setbacks in Communist China's plans to produce modern aircraft of original design. It was not until July,

1967, that the People's Republic of China announced the reopening of the Peking Aeronautical Engineering College after it had been closed for over a year. This lost time, and the difficulty of getting a specialized school of this nature back into productive service, will severely limit Chinese Communist air power for years to come. There are some signs that the Chinese are willing to recognize this difficulty and their disparity in design quality. In spite of Chinese Communist claims that their own engineers and technicians can independently develop the equipment needed for national defense, recruiting teams for the Military Science Academy operating out of Austria and Switzerland in 1966 and 1967 offered top money and guaranteed contracts to experienced German aircraft designers, engineers, and technical personnel facing loss of work in Europe. German rocket experts newly returned from Nasser's Egypt and work on the United Arab Republic missile program were also eagerly sought for their experience and technical ability. Many of these European recruits, contemplating the militant characteristics of the Cultural Revolution and the obvious Red Guard antagonism toward foreign or foreign-educated technicians since the Russian pull-out in 1960, no doubt had reservations about employment in Communist China. The over-all success of the recruiting program, however, cannot help but aid the development and production of modern aircraft and rockets of unique design in Communist China.

Communist Chinese developments in the field of military rocketry and guided missiles do not seem to have been adversely affected by the Cultural Revolution and have been advancing rapidly. A complete intermediate-range ballistic-missile complex has been constructed near Nagchu Dzong in western China, and when it is fully operational, all the border states from Afghanistan to Viet-Nam and most of the Russian defensive positions along the volatile Sino-Soviet frontier will be within range of Chinese IRBM's. On October 27, 1966, the

Chinese Communists announced the successful testing of a guided missile with a nuclear warhead. The most alarming surprise came the following summer, when Communist China exploded a hydrogen bomb (June 17, 1967), for experts had estimated that it would be years before the Chinese would be able to accomplish such a task. The fallout study indicated that the H-bomb had been released from an aircraft at high altitude. Experience with the PRC has forcefully demonstrated that today's test setup is tomorrow's military system, and the Red Chinese may therefore already have a delivery method for their new bomb. Educated opinion now places Chinese thermonuclear development ahead of that of the French, giving the PRC superpower military capability. With multimegaton-bomb development virtually racing forward on the mainland, supported by broad aircraft and intercontinental-missile delivery-system programs, both the Soviet Union and the United States will be faced with drastic revisions in their defense requirements in relation to Communist Chinese air power and missiles in the coming years.

Communist China is now concluding its frequently delayed but never terminated Five-Year Plan for the technological advancement of the PLAAF, with the program due to end in the early 1970's. Much has changed since the program was initiated, and there is little question that the results will be far different from those originally projected. Mao Tse-tung cannot long remain the political as well as the emotional leader of Communist China. Poor health, or death, must soon claim him. The status of Chinese air power will be directly tied to policies of the government that succeeds his. A strong military clique would, no doubt, attempt to increase the strength of the PLAAF and the PLANAF, while a moderate government might well find little need for such a military arm, except to suppress internal conflicts. A return to provincial government and warlordism would inevitably result in complete dissipation of the Chinese air force, while a continuation of the present form of government and control would provide the air force with its best opportunity to pursue its stated goals of self-sufficiency and planned growth.

It is still possible for the PLAAF to meet these goals. In spite of the complications of political unrest and widespread civil disturbances, the Chinese air strength both improved and increased in 1967. Estimates developed by the Institute for Strategic Studies in London place the 1968 PLAAF personnel strength at 100,000 men and the aircraft strength at 2,500 fighters, bombers, transports, trainers, helicopters and other aircraft types. This is an increase of 200 aircraft over the 1967 estimates, with the bulk of the increase made up of new aircraft types, such as the Shenyang MiG-21 fighters and their variants. The Institute reports that the air defense system, for many years providing only a thin-skinned defense of coastal areas, is gradually being expanded to protect the interior, with control being provided by ground radar and an interceptor force increasingly made up of MiG-19 and MiG-21 fighter types formed into ten to twelve fighter regiments. In addition, there are an estimated eight to ten transport regiments and an unknown number of bomber regiments, the latter constantly being reduced in number and still flying the few service-weary IL-28 and TU-4 bombers that remain from an earlier decade.

The PLANAF provides an additional 15,000 personnel and 500 shore-based aircraft to add to the strength of Chinese air power, although the estimates for 1968 suggest that no important changes have taken place in the PLANAF during the previous year. The PLANAF appears to suffer from low allocation priorities, for the 1968 estimates still include a substantial number of MiG-15bis and MiG-17 fighters in spite of the navy's participation in the integrated Chinese air defense system.

Its total 1968 military air strength of approximately 115,000 personnel and 3,000 aircraft makes Communist China the third strongest aerial power in the world, with Great Britain placing fourth with an estimated strength of 120,000 personnel and 600 combat aircraft. But Chinese air power is far behind the massive air strengths maintained by the United States, ranking first, and the Soviet Union, which ranks second with an estimated personnel strength in 1968 of over half a million men as well as approximately 10,250 combat aircraft, making it well over three times the size of the PLAAF, and the Soviet Navy adds a further 870 aircraft to the Russian air strength. The Soviet Air Force alone has about 3,700 fighters assigned to ten to twelve fighter regiments, about the same number of regiments fielded by the PLAAF, which indicates that the Chinese units are at short strength. The Soviets also maintain from eight to ten transport regiments, again matching in number the units maintained by the PLAAF, although the equipment of the Chinese regiments suffers greatly by comparison.

Compared to the air power of the United States in the Far East, the Communist Chinese Air Force is superior in number of aircraft but inferior in equipment. The 1968 USAF Tactical Air Force, with an estimated 75,000 men and 2,700 aircraft in Europe and the Far East, has a substantial portion of its strength stationed around the Chinese perimeter assigned to the Pacific Air Forces (PACAF). Made up of the 5th Air Force, stationed in Japan, Korea, and Okinawa; the 7th Air Force, in South Vietnam and Thailand; and the 13th Air Force, in the Philippines, Taiwan, and Thailand, the PACAF stands ready to counter any warlike acts by the air force of Communist China. This USAF tactical air strength is backed up by the United States Navy's carrier forces assigned to the 7th Fleet in the western Pacific, the Marine Corps Air Wings assigned to the Far East, and the additional substantial backing of the Strategic Air Command (SAC) and the Military Airlift Command (MAC), which can be brought into action as needed.

Other non-Communist air forces in the Far East provide further restraints on unlimited use of power on the part of the PLAAF and the PLANAF. The Nationalist Chinese Air Force on Taiwan, the only force currently engaging in combat with the Communist Chinese, has an estimated 1968 strength of 435 combat aircraft. Thailand has an additional 145 combat aircraft, while Japan has 570 and South Korea has 200. Facing either camp, be it the Soviet Union or the United States and its allied air forces, during the foreseeable future, the Chinese will be matched in numbers along their own borders and are overmatched in capability.

The future of Communist Chinese air power over the next decade is difficult to assess. If a judgment is to be made solely on the basis of physical and human potential at hand, there is little doubt that the PRC can solidify its position as one of the world's major military aerial powers. The material resources of the mainland, coupled with an ever growing industrial base, ensure the meeting of requirements for a commanding aircraft and aerospace industry. But the Chinese need more than their fair share of good fortune to accomplish their goals. Unless the PRC can dramatically increase its productive capacity, the current program of advancement, while providing substantial improvements over existing PLAAF stocks, merely brings some elements of the air force in closer parity with the military forces it opposes. It will take years to completely re-equip and retrain the PLAAF at the established change-over rates, and, by that time, current models, as well as China's new follow-on fighter and bomber types, will be increasingly obsolescent in relation to contemporary world standards. It is entirely possible that by the time the Chinese catch up with their adversaries the whole concept of airpower, as well as the corresponding aircraft forms and uses, will have greatly changed. The Chinese learning process would have to begin anew unless they themselves became the design leaders. Such a possibility seems very remote.

An additional limiting factor is deeply rooted in the very program designed to increase China's strength in the air. On the basis of the PLAAF growth pattern of 1967, it is obvious that China's primary current objective is to enhance its defense capability. Both the PLAAF and the PLANAF face an enormous need for new equipment, with many obsolete aircraft still remaining in first-line service. The past decade of hardship, during which the Chinese had to learn how to pull themselves out of their difficulties in the air, has left its mark etched deeply. China's isolation and the growing number of quarrels with its neighbors have created the need for a substantial defensive-aircraft inventory, yet all of these aircraft must come out of a nationalized aircraft industry subject to a single authority. There is no competitive industrial situation like those of private industry in the United States and the various aircraft-design bureaus in the Soviet Union. The Chinese are virtually forced to accept whatever design projects their most persuasive aeronautical engineers develop, with all of the inherent dangers of faulty design or lengthy debugging incorporated in the decision. Such production acceptance without recourse can lead to costly mistakes, which only superhuman engineering and phenomenal luck can prevent. Assuming that China is fortunate enough to avoid these pitfalls, the dynamics of immediate need will place another hurdle in the way of PLAAF improvement and growth. The inevitable result of a successful airframe development will be a dedication to production rather than further investigation. While Chinese production may set domestic records in the coming years, the result will actually be a second-rate air force with the false appearance of being a world leader based on the number of aircraft in service.

A historical judgment on the future of the PLAAF provides an even bleaker picture. The air force of Communist China has always been the first military arm to reflect the state of China's economy and the strength of its military establishment. Air power, owing to its sophistication and constant need of trained manpower, supplies, and maintenance, cannot long be sustained by an unsteady government or by a nation cut off from the outside world. With every ebb of Communist Chinese fortunes, the air force will feel the pinch in larger measure, as it so often has in the past. The lengthy cycles of weakness in the air after the initial establishment of the first Communist aviation units in 1926, the secretive Soviet aid in 1937, the creation of the PLAAF with massive Russian aid in 1949 and the early 1950's, and, finally, the abrupt cut in Russian aid in 1960 suggest that a period of decline is in the offing after the current rebirth of the PLAAF. PLAAF strength has never been able to sustain itself, and periods of strength and vitality have always been followed by years of decay and frustration. The current program aimed at attaining self-sufficiency may be China's only hope of breaking this pattern of history. Other alternatives are open, such as an open China freely dealing with other industrial nations, or an alteration in the basic plan, such as a total dedication to defensive and offensive missiles of original design, as has been so often claimed in the Western press but is yet to be conclusively demonstrated by the Chinese.

When Mao Tse-tung, flushed with his victory on the mainland, promised the 662 delegates of the new government of his young nation a powerful air force at the opening session of the Chinese People's Political Consultative Conference at Peking on September 21, 1949, he set the pattern for the future of Communist China in the air. Ten days later, when the People's Republic of China became an official reality, the creation of an air force was specifically incorporated in the constitution. Communist Chinese air power was thereby mated to the state as a permanent benchmark of China's progress. The fortunes of Communist China as a nation will forever be mirrored and magnified to the world by the strength or weakness of its air power.

PART TWO

THE AIRCRAFT OF COMMUNIST CHINA

INTRODUCTION

From the early "have not" days of Red Chinese aviation to the present, the aircraft of Communist China have come from widely diverse sources. The major suppliers have been the Soviet Union and, by accidents of history, Imperial Japan and the United States. Other states, such as the post–World War II Eastern European Soviet-bloc nations, have contributed to the Red Chinese aircraft inventory, creating a heterogeny of aircraft types.

This hodge-podge of aeronautical hardware has also caused Communist China training, supply, and operational problems that are not faced by other major air powers. The maintenance of a staggering variety of power plants and airframes, requiring both metric and English Standard parts and tooling, plus the many languages contained in the aircraft and their manuals, forced the Chinese Communists to take shortcuts and virtually create their own instructions and procedures. We can understand this problem if we study the various aircraft types used by the Chinese Communists over the years. It has really been only in the last decade that the Red Chinese have established order in this area. This is largely because most of their aircraft are now produced domestically, with the subsequent advantages of coordinated manuals, maintenance instruction, and crew-training procedures.

Many of the aircraft flown by the Communist Chinese in the more than forty years of their history are virtually unknown throughout most of the world. Even dedicated aviation historians, more often than not trained in other national disciplines, are frequently confused by the more obscure Russian, Japanese, and domestic Chinese aircraft types. The following glossary of Communist Chinese aircraft is intended to provide the interested aviation buff and general reader with brief insights into the development, history, and use of both the indigenous and the foreign designs flown by the Chinese Communists over the years, including those currently in use on the mainland. The purpose is to add a measure of historical significance to our account of the employment of these aircraft by the People's Republic of China. The aircraft are listed alphabetically by producer and then broken down in the order in which they are introduced in the text. This is generally in the chronological order of their introduction and use by Communist China.

Aero Super-Aero 45 Light Transport

Although it does not necessarily follow that an aircraft that looks good must therefore be good, there are notable examples where this is true. When the advanced Super-Aero 45 development of a successful light-twin low-wing monoplane entered production in 1948 at the Aeronautical Enterprises of the Czechoslovak Socialist Republic, known as Aero for short, interest in the type went far beyond the borders of this Central European nation. The Super-Aero 45 went on to become one of the world's most popular aircraft in its class, and in the ten years that it was offered for sale it was ultimately exported to more than twenty-five countries. A very attractive aircraft with a streamlined glazed-nose fuselage, the Super-Aero 45 was powered by two underslung and inverted Walter Minor 4-III inline engines of 105 hp each. The landing gear retracts into the engine nacelles, giving the aircraft a very racy appearance in the air. Largest export customer for the Super-Aero 45 was the Soviet Union. In a rare reversal of Soviet trade policy—the Soviet Union generally traded manufactured products for raw materials or soft goods—a total of 230 of the Czech aircraft were imported for use on Aeroflot, the Russian national airline. The ability to operate the Super-Aero 45 out of rough or grassy airstrips led to Chinese Communist interest in the aircraft, and a multi-aircraft sale was arranged between Omnipol in Prague, the Czechoslovakian national export trading organization, and Technoimport Peking, the Red Chinese counterpart. The Super-Aero 45 has a span of 40 ft. 4¾ in. and is 24 ft. 9½ in. long. It carries a crew of two and up to three passengers at a top speed of 175 mph. The Red Chinese later expressed interest in the Aero L-29 Delfin, a two-seat jet trainer type also purchased by the Soviet Union in 1963, but it is not known if any sales to Communist China have resulted.

Amoy Naval Establishment Pan Primary Trainer

At the time of the Fukien Rebellion, in 1933, the Nineteenth Route Army confiscated the few available Amoy Pan primary trainers then used by the Nationalist Chinese Navy, and took over the Amoy Naval Establishment Aircraft Factory. This was the only producing aircraft plant then in operation in China. The Pan trainer was a a two-seat light biplane modeled after the British Avro Avian. This original Chinese design was created by D. S. Pan, an engineer on the factory staff, and was built with Chinese raw materials and British components. The assembly and production were under the direction of the plant manager, Naval Captain Wen Lin-tschen. One of the unique features of the design was an original form of landing gear, which was attached to the fuselage by heavy rubber bands, a system that seemed to provide the required shock absorption and work quite satisfactorily. The upper wing spanned 29 ft. 6 in., and the loaded weight was barely over 1,500 lb. Top speed was 98 mph with a 90 hp Cirrus III engine imported from England. The prospect of an aircraft-producing facility in Communist hands troubled the Nationalists, but the Fukien Rebellion ended so quickly that the Chinese Communists never got the chance to join forces with the rebels and make use of their industrial prize and produce Pan trainers for their new air arm.

Antonov AN-2 Utility Transport

At a time when the whole world was jet minded, when a swept wing seemed to indicate the practical worth of an aircraft, and when new operating-speed levels were raised with every transport aircraft prototype, the Russians proudly introduced one of their newest post–World War II aircraft—a biplane. The aviation press of the early 1950's didn't know how to describe it. Some writers thought it was an old Russian design of the war years that had never been seen before. Others saw it as an indication that Russian technology was in reality well behind that of the Western nations. Only a few tried to look at the Antonov AN-2 objectively, and reported it as an intelligent solution to a uniquely Russian problem. The AN-2 looks more appropriate in the world of today than it did when it first came out. Many of the newer utility transport types produced in America and in Europe have a middle-1930's look, since they were designed to meet today's demands of short take-off, high payload, and high reliability. The Russian AN-2, designed for the same purposes, was almost a Russian "In" joke for many years. It was the first original powered aircraft design by Oleg Konstantinovich Antonov to reach production. Antonov, well known in Russia for his sailplanes and military gliders, is most famous today for his outstanding series of turboprop AN-10, AN-12, and massive AN-22 transports. He had long been interested in the lift capabilities of aircraft as a result of his work with gliders. In 1940 he re-created the design of the German Fieseler Fi. 156 Storch, one of the first truly successful STOL (short take-off-and-landing) aircraft, so that it could be produced for the Russian Army. This project came to an abrupt halt when the invading Germans captured the Russian Storch factory. They must have been surprised to see their own aircraft being assembled on a Russian production line. In 1946, in answer to the need for a new Russian agricultural and utility type to replace the twenty-year-old Polikarpov PO-2 biplane then in use, Antonov again selected the biplane layout because its high lift capabilities enabled him to create the largest aircraft possible within the confines of a reasonable set of outside dimensions. Spanning 59 ft. 7½ in. and 42 ft. long, the AN-2 held a payload of over 3,000 lb. in a loaded weight of 11,574 lb. It carried this payload in insecticides for agricultural work or in cargo; it held ten to twelve passengers in a transport version, or fourteen fully armed troops or paratroops for the Soviet Air Force. First flown in 1947, it is perhaps the most modern biplane ever built, and has the distinction of being the only biplane produced for active military service since the end of World War II. Early models had a 630 hp Shvetsov ASh-21 radial engine, but later models used the more powerful Shvetsov ASh-62IR radial of 1,000 hp, which gave the AN-2 a top speed of 155 mph. It was one of the first post–World War II aircraft supplied to Communist China; the initial examples are believed to have arrived in China during the Korean War. The NATO code name for the AN-2 is "Colt."

Aviakhim U-1 Avrushka Primary Trainer

One of the most famous aircraft to come out of World War I was the British Avro 504K trainer, an open-cockpit two-seater biplane of wood and fabric construction with a wing span of 36 ft. A number of Avro 504K's accompanied the British Expeditionary Forces to Russia in 1918 and 1919, where they were used against the Bolsheviks in support of the White Russian counter-revolution. The Russians captured a few, copied the design, and built their own at the Red Pilot Plant at Leningrad and at Aircraft Plant No. 1 in Moscow. The aircraft was produced in the Soviet Union throughout the 1920's and into the 1930's. In Russia it was designated U-1 for Trainer Model 1, and nicknamed the Avrushka, or little Avro. Even the original French LeRhone engine of 110 hp was copied and built in Russia for the U-1 as the M.2 radial engine. The Avrushka remained the standard Russian primary trainer for many years, and some of them survived until World War II. It became one of the prime aviation exports of the Soviet Union; models were supplied to various Chinese war lords for their armies, as well as to the Kuomintang-Communist coalition in 1926.

Aviakhim R-1.M-5 Reconnaissance Bomber

Flown by Royal Air Force squadrons of the British Expeditionary Force in Russia in 1919, De Havilland DH-9A two-seat biplane day bombers were perhaps the finest aircraft in their class to come out of Great Britain in World War I. An advanced development of the earlier DH-4 series, and powered by an American-built Liberty engine of 400 hp, the DH-9A was big and powerful for its day. It had a span of 46 ft. and a top speed of around 110 mph. For a while it looked as if the White Russians, under command of General Denikin, would defeat the Bolsheviks. They had stabilized a front of over a thousand miles across southern Russia, and had captured Tsaritsyn, later known as Stalingrad and now called Volgograd. The British RAF units flying their DH-9A bombers, supported by Sopwith Camel fighters, served the White Russians. They were opposed by a few Communist aircraft, mainly German pilots flying former German World War I aircraft. When the White Russian effort collapsed, some of the British aircraft fell into Bolshevik hands. Captured examples of the DH-9A were copied by the Soviets and built at Plant No. 1, the former Dux Bicycle Factory, in Moscow. The Russian De Havilland copy remained the leading general-purpose bomber, reconnaissance, and advanced trainer type in the Soviet Union until well into the 1930's. It was produced as the R-1 and the Liberty engine was reproduced as the M-5, creating the R-1.M-5 designation. Examples delivered to the National Revolutionary Army in China in 1926 were still used by the Nationalists throughout the northern march and long after the scattered Communist armies had lost their R-1.M-5 aircraft.

Avro Avian IV Trainer

One of the most popular trainer and sport aircraft of the late 1920's and early 1930's, the Avro Avian series of two-seat open-cockpit biplanes produced in Great Britain were widely exported; over fifty of them went to various governments and war lords in China and several were ultimately taken over by the Chinese Communists. Fourteen examples of the Avro 594 Avian IV model were purchased by the Chinese Naval Air Service at Amoy, and the remainder were subsequently taken over by the Aero Squadron of the Nineteenth Route Army at the time of the Fukien Rebellion in late 1933. These examples had a span of 28 ft., were powered by a 4-cylinder Cirrus Hermes II engine of 115 hp, and weighed 1,600 lb. when fully loaded.

Avro 621 Tutor Primary Trainer

Standard primary trainer of the Nineteenth Route Army Aero Squadron, and the subsequent Fukien People's Republic Air Force in 1933, the two-seat biplane Avro 621 Tutor was one of the most actively used trainers of the early 1930's. Designed in 1929 and first flown in 1930, the Tutor was produced in Great Britain and sold throughout the world. It was adopted as the standard trainer by the RAF in England and also by Canada, South Africa, Argentina, Denmark, Ireland, Estonia, Greece, Poland, and the Nationalist and revolting Kwangsi governments of China. (The Kwangsi was a provincial government with its own military forces that opposed Chiang Kai-shek in the early 1930's. The small Kwangsi Air Force was incorporated into the Nationalist forces during the war of resistance against Japan in July, 1937.) Power for the Chinese models was the 240 hp Armstrong-Siddeley Lynx IVc, which gave the Tutor a top speed of 120 mph. Span of the Avro 621 was 34 ft.

Beriev BE-6 Flying Boat

Although the Soviet Union may at first seem to be primarily a land power, a brief look at a map will soon reveal that Russia has one of the largest coastlines of any nation on earth. As a result, maritime patrol, reconnaissance, and antisubmarine aircraft have long been important to the Soviets. The Soviets have expended design energy on floatplanes and flying boats long after most of the rest of the world ignored these types in favor of long-range land-based aircraft. The Russian recreation of the American Boeing B-29 as the Tupolev TU-4 in 1944 and 1945 not only provided the Soviets with the world's most effective strategic bomber, it also gave them the most advanced piston engine of the period. Continual development of the American Wright R-3350 design by Soviet engine designer A. D. Shvetsov finally led to the 2,300 hp Shvetsov ASh-73TK engine in the post–World War II period. At the same time, the Russians developed flying boats under the direction of designer Georgi Mikhailovich Beriev, long the dean of Soviet seaplane engineering. The coming of the Cold War and the need to defend the Soviet coastline gave Beriev's work special importance, and flying-boat development proceeded apace with the availability of the new power plant that provided Beriev with just what he needed for his advanced design ideas.

When the BE-6 first flew in 1949, it gave the Russians one of the finest coastal patrol aircraft in the world. The BE-6 was large and heavy, with a span of 108 ft. 3¼ in., a crew of eight, and a loaded weight of 51,588 lb. carrying bombs, depth charges, or mines. Its large gull wing, set on a massive flying boat hull with a twin-finned tail, gave it a distinctive appearance. In spite of its size and weight, the two ASh-73TK radials gave it a top speed of over 250 mph and a range of over 3,000 miles. As newer jet-powered developments in the same basic design entered service with the Soviet Navy, a number of the Beriev BE-6 flying boats were released to the People's Republic of China after the Korean War, where they entered service with the PLANAF on coastal patrol duties. NATO gave the BE-6 the code name "Madge."

Boeing PT-17 Kaydet Primary Trainer

Backed by a rich heritage of Stearman open-cockpit biplane trainers of the 1930's, the virtually identical Lycoming powered Stearman PT-13 and the Continental powered Boeing PT-17 types were the standard primary trainers of the American Army Air Corps at the time of Pearl Harbor. They were also used by the American Naval air force for the same purposes. Chinese Nationalist Air Cadets in the United States for flight training in 1941 started to fly in the PT-17, and

148 PT-17's were shipped to China in 1942 to train more pilots in China for the war against Japan. The PT-17 was one of the last biplanes in American service. The Continental R-670-5 radial used for power was rated at 220 hp, and at full throttle the PT-17 could do 124 mph, provided everything was in prime working order. This may not sound like much in the way of speed, but in an open cockpit it seemed supersonic. Plans were under way in 1946 to produce the PT-17 in China under a Boeing license with a 6-cylinder Lycoming O-435-C inline engine of 190 hp, but the reverses of the civil war brought the project to an end. A number of PT-17's are still in use for civil and student pilot training schools on Taiwan, and an example captured by the Chinese Reds during the civil war and used by them as a primary trainer is on permanent display in Peking.

Breguet 14/400 Day Bomber and Reconnaissance

One of the most successful aircraft to be used in World War I, the French Breguet 14 series of two-seat biplanes was widely used by the French and the Americans. A Breguet 14 was the command aircraft of General Mitchell in France in 1918, and this aircraft still exists in the Smithsonian Institution in Washington, D.C. The primary models were the Breguet 14B.2 day bomber and 14A.2 reconnaissance versions. Both were similar in appearance, although the day bomber had bomb racks in the wings for attack use. With the end of the war the Breguet 14 was one of the very few wartime types to remain in production, and examples were widely exported throughout Europe, South America, and the Orient. When the major world powers elected to establish an arms embargo on China in 1922 in order to curtail warlordism and await the establishment of a responsible central government, aircraft were specifically mentioned in the embargo. France signed the agreement on one hand and then secretly supplied vast quantities of aircraft of World War I vintage to whoever would buy them on the mainland. China was flooded with obsolete and war-weary aircraft at fantastic prices in the middle 1920's from the stocks lying useless in France. The Breguet 14 began to show up all over China, and their initial sales success led to the production of a new run of seventy aircraft in 1923—known as the Breguet 14/400—specifically for Chinese service. Just about every war lord of the late 1920's or early 1930's had one or two of these bulky biplanes, and they were active participants in the numerous "summer wars." Power for the 14/400 was the 12-cylinder Lorraine-Dietrich 12Da liquid-cooled inline of 400 hp. Other examples sold in China were powered by the 300 hp Renault 12-Fox inline. The 14/400 had an upper wing span of 48 ft. 8⅞ in. and the aircraft was 29 ft. 1¼ in. long. Standard loaded weight ranged around 3,800 lb. By the time Breguet 14 production ended in France in 1926, eight years after the war, some 8,000 had been produced.

Consolidated B-24J Liberator Heavy Bomber

One of the three important heavy and strategic bombers of the U.S. Army Air Force in World War II—along with the Boeing B-17 and B-29—the Consolidated B-24 Liberator was widely used in the Pacific, in Europe, and in Africa. A massive four-engined monoplane bomber with the wing sitting high on the shoulder of the fuselage, the B-24J version weighed in at 60,000 lb. loaded, and still had a top speed of 290 mph. For power it had four Pratt and Whitney R-1830-65 radials of 1,200 hp each. The B-24 was built in greater numbers than any other World War II American aircraft: more than 18,000 were produced. The Liberator wasn't particularly liked by its crews because it was less stable than the B-17, rattled a great deal in flight, and had bomb-bay doors that had an annoying tendency to get stuck or to lock in the open position. A Nationalist Chinese squadron of B-24J bombers was trained at Pueblo Army Air Base in Colorado in 1944, but did not arrive in China with its aircraft until after the Pacific War had ended. Postwar American policy kept the Nationalists from using this force against the Communists until very late in the civil war. It would appear that the Communist Chinese claim that their first four-engine bomber was delivered to them by a defecting Nationalist pilot in June, 1946, was an honest one, for Captain Liu Shan-pan, the pilot of the B-24J that landed at Yenan, was discovered ten years later to be a deputy director of the training department of the PLAAF.

Curtiss Export Hawk II Fighter

Hottest fighter plane on the export market in the early 1930's, the American single-seat Curtiss Model 65 "Export" Hawk II biplane was virtually identical to the U.S. Navy's best carrier fighter of 1932, the Curtiss F11C-2, called the Goshawk in American service. With its 710 hp Wright Cyclone R-1820F-3 radial engine it could do over 200 mph. Put up for export sales at a time when the United States was in the throes of its greatest depression, the modern Hawk II attracted customers from all over the world. China became the best Hawk II customer. The original China demonstrator was shipped to the Orient on March 7, 1933, at a price of one dollar on the books. Jimmy Doolittle, later General Doolittle of Tokyo raid fame in World War II, demonstrated the Curtiss fighter at Shanghai, Nanking, Hangchow, and Canton. The result was an order for fifty aircraft from the Nationalist Government at an average cost of about $12,500 per aircraft. This was big money in the early 1930's, and a boon to the American aircraft industry during very dark days. As an element of the Nationalist armies, the Nineteenth Route Army Aero Squadron received part of the shipment. The Hawk II became the leading fighter plane of the Fukien People's Republic Air

PART ONE

Chinese Air Force—Taiwan

A Russian-built Aviakhim R-1.M-5 reconnaissance bomber (copy of the World War I De Havilland DH9A) in the 1926 Northern March, bearing the old Kuomintang insignia.

These Curtiss Export Hawk II fighters were part of the Nineteenth Route Army, which revolted against the Nationalists in November, 1933.

Acme Newspictures, Inc.

Jiefangjun Pao

The Lenin, a Vought V-65-C1 Corsair, which was the second Communist Chinese combat aircraft.

Jiefangjun Pao

Russian-built Polikarpov R-5 general-purpose biplanes of the Red Army air force at Yenan in 1937.

Manshu Ki.79 Type 2 (foreground) and Tachikawa Ki.55 Type 99 advance trainers (at rear), seen at Mukden airfield in August, 1945, after they were seized by the Russians.

Denys Voaden

Jiefangjun Pao

Red Chinese troops replacing propeller on captured Japanese Tachikawa Ki.55 Type 99 advance trainer.

Koku Fan

Former Japanese Tachikawa Ki.55 Type 99 advance trainer on display at People's Revolution Military Museum, Peking, in 1966.

Peter M. Bowers

Captured Japanese Kawasaki Ki.61-I Type 3 fighter, photographed in Peking in Nationalist markings in November, 1945.

Peter M. Bowers

The Nakajima Ki.84 Type 4 fighter, one of the two types of aircraft forming the backbone of the early Red Chinese fighter units.

David W. Lucabaugh

Captured Nakajima Ki. 44 Type 2 fighter, used by the Communists early in the civil war.

Jiefangjun Pao

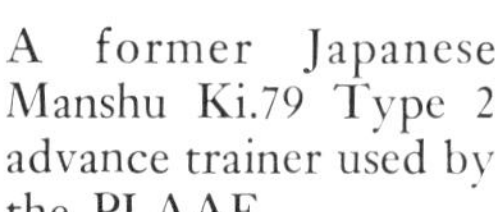

A former Japanese Manshu Ki.79 Type 2 advance trainer used by the PLAAF.

David W. Lucabaugh

Former Japanese Army Kawasaki Ki.48-IIb light bombers, manned by PLAAF pilots beginning in 1946.

A former Japanese Tachikawa Ki.54a two-engine advance trainer.

Jiefangjun Pao

AP Newsfeatures

Chou En-lai, Chu Teh, and Mao Tse-tung at the Yenan airfield early in 1947. In the background, the American Douglas C-47 type transport put at Mao's disposal by the United States.

Nationalist P-51D fighters, which were captured in whole squadrons by the Chinese Communists during the summer of 1949.

David W. Lucabaugh

North American B-25H Mitchells were captured by the Communists along with the bulk of the Nationalist Air Force in 1949.

Eastfoto

North American P-51D Mustangs, photographed in 1951. They were the standard PLAAF fighter in the closing months of the Chinese civil war.

A Soviet instructor with Chinese pilots, 1950.

Jiefangjun Huabao

U.S. Defense Department

The Ilyushin IL-10 ground attack plane, over one hundred of which were delivered by the Soviets in 1950 during the complete reorganization of the Communist Chinese Army and Air Force.

Sekai no Kokuki

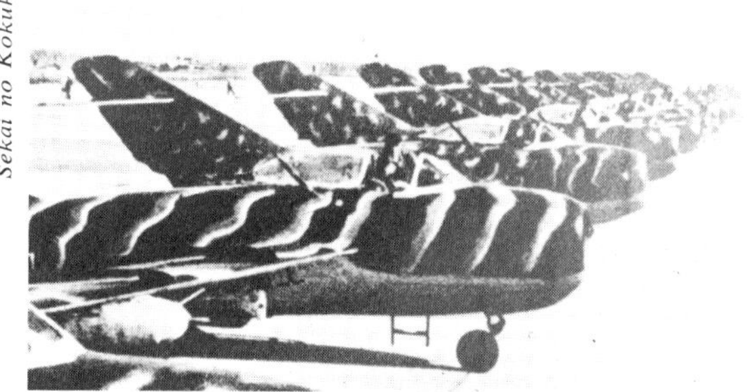

The first Chinese Mikoyan MiG-15 fighters, which began to show up over Korea in late November, 1950.

A Chinese Communist pilot seen ejecting from his MiG-15 fighter over North Korea in May, 1952, after his aircraft was incapacitated by an American F-86 Sabre pilot.

USAF gun camera close-up recording a Red Chinese MiG "kill" over North Korea.

USAF

Li Han, the first PLAAF pilot to down an American aircraft over Korea.

Jiefangjun Pao

Chang Chi-wei, credited with downing the American "ace," Major George Davis, on February 10, 1952. Stars on the side of the MiG denote "kills" of USAF aircraft.

A Russian-supplied Tupolev TU-2 attack bomber, of the type actively used on ground support missions and against the Korean offshore islands.

Koku Fan

Jiefangjun Pao

Lt. Colonel Wang Hai (center), another Communist Chinese Korean War "ace," commanded the First Interceptor Regiment of the PLAAF.

Red Chinese Mig-15 fighters, photographed at a Manchurian base near the North Korean border.

Sekai no Kokuki

Eastfoto

Women pilots, seen here in February, 1953, flew former Nationalist Douglas C-47 transports in support of the PLAAF.

U.S. Defense Department

PLAAF aircraft insignia: red star and horizontal bars outlined in golden yellow; the white characters in the center of the star signify the Gate of Heavenly Peace in Peking.

Russian Yakovlev YAK-11 advance trainer. These aircraft were supplied during and after the Korean War to replace dated Japanese trainers used by the PLAAF.

Jiefangjun Huabao

PART TWO

Liang Yen-wen

The Red Chinese Mikoyan MiG-17 fighter photographed over Shanghai in 1956.

PLA troops practicing tactics, with the help of a Russian-built Mil MI-4 helicopter.

U.S. Defense Department

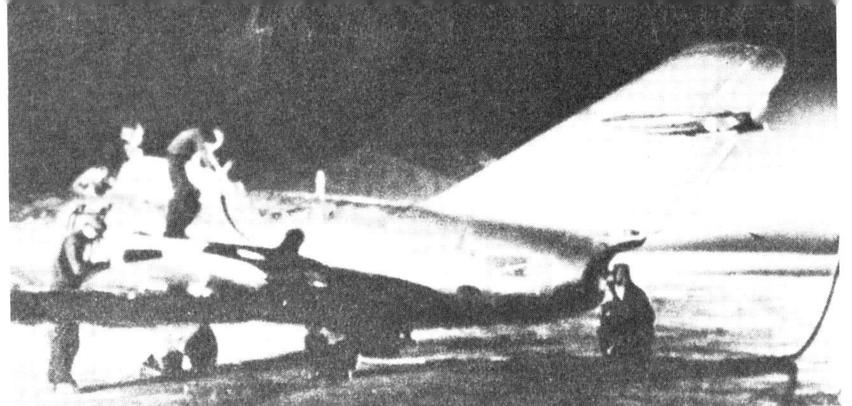

A Russian-built MiG-15bis of the People's Liberation Army Naval Air Force assigned to coastal area defense facing the Taiwan Strait.

Jiefangjun Pao

Another view of a PLANAF MiG-15bis fighter.

A Russian YAK-11 at the Air Academy at Sian in May, 1955.

Jiefangjun Huabao

China Pictorial

A China-Salamandra glider, an adaptation of the Polish single-seat SZD Salamandra.

A Chinese-built glider used in pilot training.

The Polish SZD-8 Jaskolka, a sailplane flown by the PLAAF.

China Pictorial

Pilot Chang Tsai-wen with her Chinese-built PLAAF SZD Jaskolka sailplane.

Sekai no Kokuki

Production of a Chinese version of the Russian-designed Mikoyan MiG-17 fighter at the National Aircraft Factory, Shenyang, Manchuria.

Jiefangjun Pao

A Chinese-produced MiG-17.

Soviet Sailor

Another view of the Chinese-built Shenyang MiG-17 fighter.

China Pictorial

A Shenyang YAK-18 primary trainer monoplane, photographed at a model airplane meet at Peking in the spring of 1962.

License-built Yakovlev YAK-18 primary trainers of Russian design, produced at the Chinese National Aircraft Factory.

China Pictorial

Sekai no Kokuki

The first civil aircraft built in Communist China, a license-built version of the Russian Antonov AN-2 utility biplane (December, 1957).

China Pictorial

A Shenyang AN-2 Fong Shou No. 2 biplane of the Special Flight Group, used for crop spraying and mosquito control.

Mil MI-4 helicopter produced at Shenyang.

Eastfoto

Jiefangjun Huabao

Training of PLA assault forces in the late 1950's, with Shenyang-produced MI-4 Whirlwind helicopters.

The Feilung No. 1, a twin-float Red Chinese version of the Russian Yakovlev YAK-12.

Eastfoto

The Bei-Jing No. 1, a feeder-line transport based on the Russian Yakovlev YAK-16, shown under construction at the Peking Aeronautical Engineering College in the summer of 1958.

Sekai no Kokuki

Prototype Bei-Jing No. 1, first flown on September 24, 1958.

The Sungari No. 1, built at Harbin engineering works in Manchuria, a license-built version of the Czech Super-Aero 45, used by both the civilian line and the PLAAF.

Sekai no Kokuki

Eastfoto

Russian-built Lisunov LI-2 cub passenger transport on the runway at Kashgar in the Gobi Desert. This CAAC transport took part in the China-USSR passenger service opened up on January 1, 1955.

Wide World Photos

A Curtiss C-46 Commando, one of seventy-one Nationalist civil aircraft briefly owned by the Communists at Hong Kong in February, 1950, receives its People's Republic flag fin marking.

China Pictorial

The transport of livestock, such as pedigree horses, is promoted by the CAAC along with other services besides passenger transport. Aircraft used here is a Lisunov LI-2 (Russian-built version of the Douglas DC-3).

Eastfoto

A ceremony at Taiyuan on July 11, 1958, marks the opening of a new CAAC feeder air route. The aircraft is a Shenyang AN-2 Fong Shou No. 2.

The Super-Aero 45 monoplane, another regular CAAC transport for feeder routes.

Omnipol

A Russian-built Ilyushin IL-14M transport was used for the first flight on the CAAC route between Kunming, in China and Burma, April 11, 1956.

Peter Keating

An East German VEB-Dresden IL-14P transport (a license-built version of the Ilyushin design), used in CAAC international flights. Photographed at Rangoon in the spring of 1962.

China Pictorial

One of the large turboprop powered Ilyushin IL-18 transports obtained from Russia, the pride of the CAAC routes.

CAAC transports carry the airline name in characters, and the colorful CAAC insignia on the forward fuselage (red central star, light-blue wings, and black smaller stars).

Peter Keating

Eastfoto

Pilot officer Pao Shou-ken, an instructor like many other veterans of the air battles of the Quemoy Straits, trains a class in air-to-air combat.

Red Chinese pilot Kao Chang-chi, credited with downing a Nationalist McDonnell RF-101 Voodoo reconnaissance plane over the mainland in the spring of 1965.

Four Lockheed U-2 reconnaissance planes of the Nationalist Air Force, downed over Red China and placed on display at the People's Revolution Military Museum in Peking, 1965.

Aero Fan

PART THREE

Eastfoto

Russian IL-28 medium bombers (Korean War vintage), seen with the Eighth Air Division of the PLAAF during maneuvers, summer, 1964.

Several PLAAF officers photographed in the distinctive air force uniform that was abolished along with military ranks in Communist Chinese military forces on June 1, 1965.

Jiefangjun Pao

Sekai no Kokuki

Outdated Shenyang MiG-17 fighters, seen on a Red Chinese airfield, are still used by many PLAAF regiments.

Eastfoto

The Ilyushin IL-28 is the standard medium bomber of the PLAAF, and the only one available in quantity.

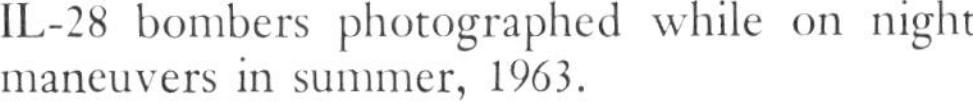

IL-28 bombers photographed while on night maneuvers in summer, 1963.

China Pictorial

Modified examples of the IL-28 designed to carry torpedoes, still used by naval PLANAF regiments.

PLAAF mechanics and ground personnel engaged in daily maintenance of Ilyushin IL-28 bombers.

Jiefangjun Huabao

China Pictorial

The Peking-based crew on a PLAAF Ilyushin IL-14 military transport, publicized in the Chinese press for having rushed medicine to road workers in Shansi suffering from food poisoning, on February 3, 1960.

Eastfoto

Shenyang MiG-17 fighters, seen in a fighter scramble at a PLAAF base in South China in the summer of 1965.

China Pictorial

This PLAAF Shenyang Whirlwind 25 (a Chinese version of the Russian Mil MI-4) was used as an ambulance in March, 1966, following an earthquake in the Hsingtai area.

Members of the Haichian People's Commune described in the press as amateur peasant parachutists participating in "mass sport." Their aircraft is a PLAAF Shenyang AN-2.

Two Red Chinese pilots, Shao Hsi-yen (left) and Kao Yu-tsung, defected in a Shenyang AN-2 from the mainland to South Korea while on an insect-control mission.

The Shenyang AN-2 flown by Shao and Kao, displayed as part of the Nationalist Freedom Day celebration in Taipei on January 23, 1962.

Taiwan Pictorial

Red Chinese Naval Lt. Liu Cheng-sze defected to Taiwan on March 3, 1962, in this Communist Chinese Mikoyan MiG-15 bis, photographed immediately after landing.

Lt. Liu Cheng-sze's Mikoyan MiG-15bis (here seen on display in Taiwan) is typical of the obsolete equipment in wide use in Red China's air arms.

Liang Yen-wen

Koku Fan

This former Nationalist North American F-86F Sabre jet is on display at the People's Revolution Military Museum in Peking, after its pilot defected from Taiwan.

The first Red Chinese bomber to fall into non-Communist hands, this IL-28 reached Taiwan on November 11, 1965, when its pilot defected.

The IL-28 skidded off the runway at Taoyüan Air Force Base on Taiwan during its landing.

Central News Agency

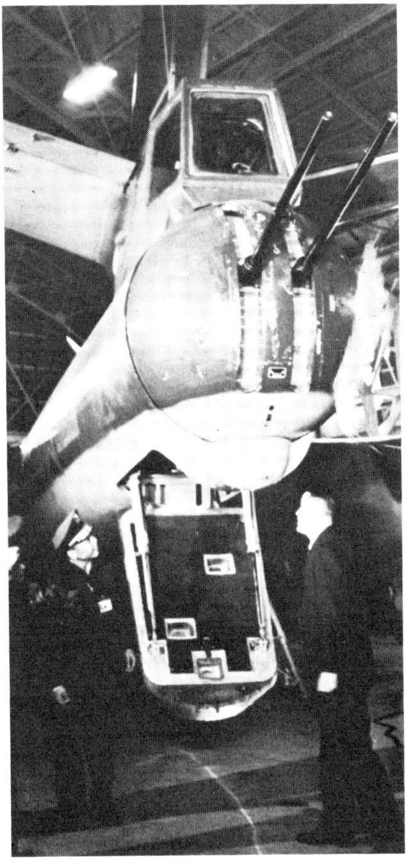

Central News Agency

Left Nacelle of the former PLAAF IL-28, which has been repaired and test flown by the Nationalists and studied by United States technical personnel.

The rear access hatch and the tail gunner position of the defected IL-28 bomber (note the flat bulletproof glazed panels).

Koku Fan

The newest fighter reaching PLAAF service is the Mikoyan MiG-21, re-created by Chinese engineers at Shenyang from Russian models.

Wide World

The wreckage of three pilotless United States Ryan BQM-34A Firebee drones, reportedly shot down by the PLA over China in 1965, were put on display at the People's Revolution Military Museum in Peking in April, 1965.

Close-up of one of the Ryan drones in Peking as it is being viewed by visitors to the Military Museum. The name Crazy Legs appears on the nose. It reportedly bore a 1964 manufacturing date.

Associated Press

Force. A number were destroyed by the Nationalists during the Fukien Rebellion, including some in air combat over Foochow. Delivery of all fifty of the aircraft of the Nationalist Order was completed by September, 1933.

Curtiss C-46 Commando Transport

When it was introduced in 1940, the 36-passenger Curtiss CW-20 civil transport was heralded as the largest twin-engine transport plane in the world. It had a span of 108 ft. and a loaded weight of over 50,000 lb., over 50 per cent more than a loaded Douglas C-47. Military requirements led to production as the C-46, and over 3,000 were produced by the end of World War II. The massive C-46 turned out to be a history-making aircraft. It formed the backbone of the aerial supply route to China, flying the "hump" of the Himalaya mountains from India to Kunming, China, to keep the embattled Chinese and the American Fourteenth Air Force supplied with fuel, arms, and supplies in spite of a Japanese sea and land blockade of the Chinese mainland. Hundreds were turned over to the Chinese for military use and sold to Chinese operators for civil use after the end of World War II, and many are still in use as military and civil types on Taiwan. Many more were taken over by the Chinese Communists during and immediately following the civil war and some are still flying on the Chinese mainland.

De Havilland DH-98 Mosquito Fighter

Claimed by many to be the outstanding British aircraft of World War II, the wooden mid-wing twin-engined De Havilland DH-98 Mosquito was produced in a plethora of bomber, fighter, and reconnaissance models. Production of the Mosquito originated in England, but later in the war it was also produced in Canada, where 1,134 airframes of various models were built by October, 1945. Power plants of the Canadian versions were Packard Merlins of 1,620 hp each. The end of the war left Great Britain and Canada with surplus stocks of war goods, and the Mosquitoes began to show up in many air forces around the world. Halfway around the globe, as the tide of the fighting began to turn against the Nationalists on the Chinese mainland, government officials desperately searched for modern aircraft to use against the Communists. By late 1947 negotiations were underway between the Nationalists and Canada for 150 Mosquitoes at a price of about $10,000 each. Ultimately some 250 were shipped to China, where they served in the civil war, with their last stand being made in round-the-clock Mosquito raids against the mainland from the last remaining Nationalist bases on the island of Hainan in February, 1950. A substantial number of the Canadian Mosquito bombers and fighters fell into Red Chinese hands.

De Havilland-Canada DHC-2 Beaver Utility Transport

Originally established in 1928 as a Commonwealth facility to sell and service British De Havilland designs in Canada, the De Havilland Aircraft Company of Canada, Ltd., became a major aircraft-producing establishment during World War II. After the war De Havilland-Canada began to produce its own designs under the DHC designation. The first aircraft to be produced was the DHC-1 Chipmunk, a two-seat primary trainer monoplane that received wide international acceptance. Following this success, the DHC-2 Beaver was built in 1947 and became the first in a long line of Canadian STOL aircraft, making De Havilland-Canada one of the most experienced producers of short-take-off-and-landing aircraft. A high-wing cabin monoplane looking like an overgrown light plane, the DHC-2 is powered by a 9-cylinder American Pratt and Whitney R-985 Wasp Jr. radial engine of 450 hp. Wing span is 48 ft. and length 30 ft. 3 in. Although somewhat smaller, carrying a pilot and up to seven paratroopers or passengers, the DHC-2 utility transport is in the same class as the Russian Antonov AN-2, a type also produced in Red China at the National Aircraft Factory as the Fong-Shou No. 2. Over 1,500 DHC-2 aircraft were built in Canada, and the type was even accepted by the United States Army as the L-20A transport. It was also sold throughout the world, and examples were delivered to the Indian Air Force. The Red Chinese held a captured example long enough to study it carefully and expose their engineers to this outstanding Western STOL type.

Douglas 0-2MC4 Marx General Purpose

When Chiang Kai-shek's government at Nanking began negotiations to buy American Douglas 0-2MC4 light bomber and general purpose biplanes in 1930, loud cries went up from rival Chinese governments, including the Communists, about Chiang's warlike intentions. An export version of the old U.S. Air Corps 0-2 Observation series, the two-seat open-cockpit 0-2MC4 gave the Nationalists the strongest and most modern air arm on the mainland. Additional examples were ordered in the early 1930's until this type became the most common aircraft used by the Nationalists. They called it the Big Douglas to differentiate it from a lower-powered trainer version also in Chinese use called the Small Douglas. The Big Douglas, in evidence wherever Nationalist power was felt, became feared by the forces opposed to Chiang. It was the standard reconnaissance bomber type used against the Communists and constantly harassed them along the route of the Long March. The power plant was a Pratt and Whitney Hornet 9-cylinder air-cooled radial of 525 hp, an aircraft engine particularly easy to maintain. The 0-2MC4 could carry a useful load of 1,750 lb. and had a top speed of 145 mph. Although the single captured Communist example was soon lost, a great number of the

Nationalist examples remained in use with Chiang Kai-shek's forces for many years, and some even took part in the fighting in China against the Japanese as late as 1944.

Douglas C-47 Dakota Transport

Probably the best-known aircraft in the world even today, the U.S. Army Air Force Douglas C-47 was a military development of the middle-1930's Douglas DC-3 airliner, the most widely used transport in the world both before and immediately after World War II. The DC-3/C-47 served on both sides of the war in great numbers—it was built in a wide variety of models for the Air Force and the Navy of the United States during the war years and was also produced under license both in the Soviet Union and in Japan—and was available by the hundreds when the war was over. A fabulous number of these old "Gooneybirds" and their 1200 hp Pratt and Whitney engines are still doing their jobs well without fanfare. After flying in daily service for over thirty years, the DC-3/C-47 seems ready to go on for another thirty. Even to this day some of the new feeder-line transport aircraft are described as "DC-3 replacements." Of the many wartime transport models produced, the C-47 Skytrain and C-53 Skytrooper models were produced in the greatest numbers, although the official names never took hold. All of the military DC-3 models were called the Dakota by the British, and that name seems to have stuck. The specific model given to Mao Tse-tung by the United States has not been reported, although it appears to have been a transport conversion of a C-47 with practical, if not overly comfortable, seating for its new owners. The DC-3/C-47 series had a wing span of 95 ft. and were 64 ft. 6 in. long. Loaded weight was rated at about 24,400 lb., but there are many stories about these aircraft flying with fantastic overloads. Top speed was barely over 180 mph at about 6,500 ft. and dropped off after that the higher the aircraft climbed, up to its service ceiling of 20,800 ft.

Feilung Hiryu No. 1 General Purpose Seaplane

The Communist Chinese technique of giving different names to YAK-12 developments produced at various manufacturing plants on the mainland may convince the Chinese that their homeland is producing a wide variety of different aircraft, but Western observers are able to see these aircraft for what they really are: copies of the Russian type modified for special duties. The Feilung Hiryu, or Flying Dragon, is virtually identical to the Shenyang Chinko No. 1, except for the twin floats in place of the Chinko's fixed landing gear. Production of the Hiryu was assigned to the Feilung Machine Building Works in Shanghai, an old factory building left over from Nationalist days, and the prototype was rolled out—or, more correctly, floated out—on October 11, 1958. Once again the Chinese Communists made rash claims, stating that the Hiryu was China's first hydroplane, whereas seaplanes of original Chinese design had been produced at virtually the same location, perhaps even in the same plant, since 1931. Even before that, as early as 1916, Chinese seaplanes had been produced by the Naval Air Establishment set up during the days of the old Chinese Republic at Foochow. The Hiryu No. 1, as the aircraft was designated, was built for use in south China on the canal networks reaching deep into the rural areas. It carried four passengers, including its pilot, and made use of the same Chinese M-11FR power plant mounted on the Chinko No. 1. Quantity production of the Hiryu No. 1 was initiated in 1959.

Handley Page HPR7 Herald Medium Transport

The sale of six Vickers Viscount 843 transports to Red China in 1961 opened the doors to additional aircraft business between Britain and mainland China, the twin-Dart turboprop-powered high-wing Handley Page Herald transport receiving particular attention. The Herald was conceived as a short-range transport in the class of the Fokker/Fairchild Friendships and the Russian Antonov AN-24, all of which had a similar design format. The Herald made its first flight on March 11, 1958, and was regarded as a good export item by the British aircraft industry. But sales became tougher as foreign competition in this transport class increased, and as a result the Herald did not attract the wide sales enjoyed by the earlier Vickers Viscount. A special overseas demonstration flight with Prince Philip, Duke of Edinburgh, and British European Airways Captain S. J. Nicolle at the controls was made early in 1962 to attract Herald sales, and a growing list of orders ultimately resulted. It wasn't until June, 1965, that the Handley Page Company was able to announce the conclusion of a sale of two Heralds to Red China for about $3 million, and when the announcement was made the firm carefully noted that the transports were for Chinese "internal service and charter operations." American concern over the sale of these aircraft to Red China was based on the fact that such aircraft would build up the Chinese Communist limited war potential, by providing China with modern aircraft that could airlift supplies to Laos or Viet-Nam. Additional Herald sales to Red China, or the prospect of a Chinese copy of the aircraft, make this a very real concern. In its original series 100 format the Herald has a wing span of 94 ft. 9½ in., a loaded weight of 39,000 lb., and carries forty-four passengers. The "stretched" version ordered by the Chinese carries up to 56 passengers. Power is two Rolls-Royce Dart Mark 527 turboprops of 1,910 hp, although later models mount the Dart

RDa.10-1 rated at 2,775 hp with the added boost of 740 lb. of jet thrust.

Harbin Aeronautical Engineering Works
Hai Lun-Kiang No. 1 Agricultural Plane

The industrial complex of Manchuria is the very heart of the Red Chinese aircraft industry, not only for the sophisticated military types produced for the PLAAF but also for the agricultural and commercial types so desperately needed on the mainland. The city of Harbin, located some 300 miles north of Shenyang, is also an important aircraft-producing site. The second largest plant of the former Japanese-controlled Manshu Airplane Manufacturing Company had been located there to produce aircraft engines during the war, and skilled workers remained in the area when the Communists took over. The first reported Red Chinese aircraft to be built at Harbin was the Hai Lun-Kiang No. 1, yet another Red Chinese modification of the Russian YAK-12 design with a Chinese-built M-11FR radial engine of 160 hp. This version was built in prototype form by the teachers and students of the Harbin Aeronautical Engineering College of the Harbin Polytechnic University, one of the five leading universities of its kind in Communist China. Founded in 1920 as a school for railroad technicians, it was reorganized in 1949 along Russian lines to include a five-year "learn and work" undergraduate program as well as a graduate school conducted entirely in Russian. Among the ten technical colleges within the university was the aeronautical engineering school, whose faculty included several Soviet instructors, who were no doubt influential in adapting the YAK-12 design for development at the Harbin Aeronautical Engineering College. A completely new fuselage and vertical tail on the Harbin Hai Lun-Kiang No. 1 have changed the appearance of the basic YAK-12 design and provide room for three passengers including pilot, instead of the seating for four in the other versions. A large cargo area has been added for light transport or crop-spraying duties, and a longer and higher landing gear makes the aircraft sit higher on the ground. The redesign has improved the STOL characteristics of the design, permitting the Hai Lun-Kiang No. 1 to take off and land in less than 250 ft. Useful load is 484 lb. and top speed is just short of 100 mph.

Harbin Aeronautical Engineering Works
Sungari No. 1 Light Transport

Examples of the Czech Super-Aero 45 had hardly entered service in Red China before a Communist Chinese redesign was undertaken, in 1958, by the students and instructors of the Harbin Aeronautical Engineering College, a technical school located on the same site as an aircraft production facility. The design work and prototype construction of the twin-engine low-wing monoplane type were reportedly accomplished in eighty-one days, and the aircraft was called Sungari No. 1, the name of a large Manchurian river running through the city of Harbin. The Communist Chinese adaptation lost some of the attractive lines of the original Super-Aero 45, particularly in the fuselage. The Red Chinese engineers had elected to modify the glazed nose of the Czech version and replace it with a more conventional pilot's enclosure with a windscreen set-back. Production of the Sungari No. 1 was immediately initiated at the Harbin Aeronautical Engineering Works, the plant in which the aeronautical students applied their classroom theories to production in typical Communist Chinese style. The first of the new Chinese models reached CAAC service before the end of 1958. The Sungari No. 1 was placed in airline service on both feeder-line passenger and freight use. The aircraft is capable of carrying five people or 1,410 lb. of cargo. Engines for the Sungari are probably Walter Minor 4-III inlines of 105 hp imported from Czechoslovakia, since they have the same lines as those used on the original Super-Aero 45. The "learn and work" procedure that led to the Sungari No. 1 was severely criticized a few years later when Chu Wu-hua, former vice-president of Harbin Polytechnic University, and in 1961 an engineering deputy to the Third National People's Congress in Peking, spoke out against the wholly pragmatic approach to Chinese technical education. Chu stated that most students were interested only in applying their technical knowledge to actual production, and that their "narrow-mindedness" would ultimately be costly to China. Other voices soon complained that the instructors who had graduated from the Chinese technical schools in the mid-1950's had the same failing. Many of these militant students and instructors became Red Guards when all of the technical schools were closed during the aggressive days of the Cultural Revolution. Now that the schools have started to reopen, it will be enlightening to see if the theoretical or pragmatic approach to aircraft design gains the upper hand.

Ilyushin IL-2m3

At first glance the Ilyushin IL-2m3 monoplane appears to be a somewhat obscure type of Russian World War II aircraft designed and modestly produced for special needs over the Russian front. Nevertheless, the IL-2 series was produced in greater numbers than any other aircraft in the history of aviation. Not even the massive American production totals of the World War II years, which ran into the tens of thousands for a number of specific aircraft types, could match the estimated production total of 35,000 Ilyushin IL-2 ground attack bombers, or the fantastic total of between 39,000 and 40,000 IL-2's that includes the later IL-10 model. The IL-2 was the Soviet Union's most famous World War II aircraft, and although it was not a thing of beauty, it is remembered as fondly in Russia as the Mustang in the United States and the Spitfire in Great Britain. Throughout the IL-2 series the basic design format of a heavily armed and armored low-wing monoplane with an inline engine for low-level use against tactical ground targets remained unchanged, although a few radial powered prototypes were built for test purposes. The Russian pilots called these types the Ilyusha, based on the producer's name, Ilyushin, although they were generally known in the west as the Shturmovik (assault aircraft). The initial IL-2 models were single seat, and when they began to raise havoc with German armored units during the invasion of Russia in 1941, the Shturmovik captured the popular imagination and became one of the first Russian aircraft to receive worldwide recognition. Excessive losses to German fighters led to the development of a two-seat version as the IL-2m3 with provision for a rear gunner to protect the tail. The IL-2m3 took part in the pivotal battle of Stalingrad. From the end of 1943 until the war in Europe was over, thousands of them literally swarmed over the German front lines. The aircraft mounted a fantastic array of weapons for its day, normally carrying eight 82 mm anti-tank rockets, two forward 23 mm cannon, a rear 12.7 mm machine gun, and 882 lb. of bombs. Its 1,750 hp AM-38f liquid-cooled inline engine gave it a top speed of 250 mph at around 5,000 ft. Loaded weight was 12,950 lb. After the war the IL-2m3 was supplied to Czechoslovakia, Poland, Yugoslavia, and the People's Republic of China. It remained in active service long enough to receive the NATO code name "Bark."

Ilyushin IL-10

Last in the Ilyushin line of Shturmovik attack planes, the two-seat IL-10 was an entirely new design based on the same concept as the IL-2m3. As such, it came out looking much like the earlier model, although the refinements were obvious after careful inspection. It was a larger and heavier aircraft and had cleaner lines. The first IL-10 attack planes entered service in the Soviet Union late in 1944, and in the closing months of the war in Europe the IL-10 was supporting Russian troops in their drive for the heart of Germany. The IL-10 was quickly distributed to the new Communist air forces that formed in Eastern Europe as the Soviet sphere of influence virtually exploded following the political vacuum created by Germany's defeat. Substantial numbers of the estimated 2,000 IL-10 aircraft produced in Russia were delivered to Rumania, Bulgaria, Yugoslavia, Poland, Hungary, Czechoslovakia, and eventually to the People's Republic of China and the Korean Democratic Republic. It was in Korea that the IL-10 was first used against American forces; examples were captured there and evaluated in the United States. Although the propeller-driven IL-10 might be considered an obsolete type to foreign eyes, it was in the same class as the old American Douglas Skyraiders in use in Viet-Nam in the 1960's and served the Chinese Communists as a counter-insurgency type long and well on the mainland. In 1950 license production of the IL-10 was undertaken as the Avia B-33 in Czechoslovakia, where an additional 2,000 were produced in the next four years. The Russians probably recommended production of the type in China, although Chinese Communist insistence on only the newest and the best in the way of aircraft would have thwarted this suggestion. Power for the IL-10 was a 12-cylinder Mikulin AM-42f liquid-cooled inline of 1975 hp, which gave it a top speed of 315 mph at sea level. Span was 45 ft. and length 40 ft., and the normal loaded weight was 13,968 lb. The armament was a marvel for its time; the Communist Chinese models of the IL-10 carried two 23 mm cannon and two 7.62 mm machine guns in the wings, a 12.7 mm machine gun in the rear cockpit, eight 82 mm rockets, and 882 lb. of bombs. The Russians long maintained their appreciation for the IL-10, for it remained in first-line Soviet service until 1956—eleven years after World War II. It served in China long after that. The IL-10 received the NATO code name "Beast."

Ilyushin IL-12 Transport

At the height of World War II the Russians recognized the need for a replacement for their Lisunov LI-2 redesign of the American DC-3 transport, the standard Russian military transport of the war years. The designer of the fabulously successful Shturmovik series, Sergei Vladimirovich Ilyushin, was given the task of creating the new transport in 1943 utilizing Shvetsov ASh-82 14-cylinder radials of 1850 hp, the power plant also selected for use on the new Lavochkin fighters then under development. The format selected by Ilyushin was virtually identical to the path taken by American designers facing the same problem of a DC-3 replacement, and when the war was over and the IL-12 produced by Ilyushin began to show up in military and civil versions in the Communist world it was given the nickname of the Russian Convair. The similarities between the independently designed postwar American Convair 240 transport and the IL-12 transport produced in Russia were so obvious that it looked almost as if one had been copied from the other, although it was impossible to tell which one came first. The Russian design actually came first, for IL-12 transports were flying in 1944. They became the standard postwar transport of the Soviet Union and were supplied in large numbers to Communist-bloc nations. Ilyushin received a Stalin Prize for his design. Military models were built as cargo and troop transports, and had gun ports at the windows. They could carry 8,800 lb. of cargo or troops. The civil versions carried from twenty-four to thirty-two passengers and quickly entered service with Aeroflot, the Russian national airline, in great numbers after the war. The IL-12 just as quickly showed up in the many new air forces and airlines operated by the Communist governments of Czechoslovakia, Poland, East Germany, and other satellite nations and during the Korean War in Communist China. Although it looked smaller, the IL-12 span of 104 ft. made it 9 ft. larger than the earlier LI-2. Many of these transports are still flying. The NATO code name is "Coach."

Ilyushin IL-14M Transport

An improvement of the Ilyushin IL-12 transport series making its appearance in 1953, the twin-engine low-wing monoplane IL-14 had more powerful 1,900 hp Shvetsov ASh-82T 14-cylinder radial engines, the addition of exhaust thrust to the nacelles, and refined lines including a large squared-off vertical tail. The new tail is the single most notable design change to the eye and thus permits easy distinction between the IL-14 and the IL-12. This distinction would otherwise be difficult, since these aircraft look very much alike. The dimensions of these two transports are virtually identical—the IL-14 has the same span as the IL-12 but is a scant one inch longer. The IL-14 also sits five inches lower to the ground with a height of 26 ft. 1 in., giving it the look of a kneeling IL-12. The IL-14's top speed is 237 mph, about 10 mph faster than the earlier transport. A wide range of models have been produced for both military and civil use, and the IL-14 became the mainstay of most of the airlines behind the Iron Curtain in the middle 1950's. Military models, including the IL-14T, were produced for cargo, infantry, paratroop and VIP use. The IL-14M was one of the later Russian-produced civil transport models, and Red China became one of the prime customers for the type. Loaded weight with twenty-four passengers was 38,030 lb. Licensed production of the IL-14 was also authorized in East Germany and Czechoslovakia, and in 1959 it was reported that Red China also considered production of the type although nothing seemed to come out of this suggestion. Either the report was in error or the Red Chinese shelved the idea as they had done with a number of other Russian aircraft types that were submitted by the Soviet Union for production in Red China. NATO code name for the IL-14 is "Crate."

Ilyushin IL-18 Transport

Looking somewhat like an overgrown IL-14 with four closely cowled streamlined engines, the Ilyushin IL-18 was the Soviet Union's most widely used turboprop transport in the middle 1960's. Its closest Western counterparts are the American Lockheed Electra and the British Vickers Vanguard. The IL-18 first flew in July, 1957, quickly entered Aeroflot service in the Soviet Union on a test basis in 1958, and was aggressively offered on the export market both within and outside of the Soviet bloc. By the end of 1958 Communist Chinese pilots, engineers, and mechanics were already being sent to Moscow for training on the IL-18, in preparation for its use by CAAC on long-range routes in mainland China. The rush to get the IL-18 into service overshadowed the very real problems faced with its four 4,015 hp Ivchenko AI-20 turboprops. Foreign operators in East Germany, Czechoslovakia, Hungary, Bulgaria, Africa, and Communist China soon discovered that the turbines broke down well before their rated overhaul due dates, and the initial Chinese CAAC IL-18s were finally returned to the Soviet Union for repair and explanation. The Russians denied the existence of the problem for some time, claiming that poor maintenance practices caused the breakdowns and high operating expenses. But not even the Russians could avoid facing the problem, for overhaul histories on the AI-20 turboprops of the examples flown by Aeroflot soon revealed that breakdowns after 300 hours were not extraordinary. Finally, late in 1963, no less a personage than Soviet Chairman Nikita S. Khrushchev, speaking in Moscow to the Communist Party Central Committee, admitted that Russian civil aviation faced serious problems with the power plants on their IL-18 transports. He confessed to a 500-hour overhaul requirement while stating that British turboprops performed well anywhere from 2,200 to 2,500 hours between overhauls. A massive attempt to overcome the faults was made on the domestic and export models of the turboprops, and while the newer AI-20 turbines still do not compare in service with their foreign competition the Russians now claim a 12,000-hour service life with two overhauls. At least one of the Red Chinese IL-18's has been outfitted as a VIP transport and, marked in PLAAF insignia, has been used on influential trips outside of the People's Republic of China, such as premier Chou En-lai's journey to Africa and Egypt in December, 1963. Other examples are said to be in PLAAF service as troop transports. In its civil role the standard IL-18v model has a crew of five and can carry from 73 to 111 passengers. Span is 122 ft. 8½ in. and length 117 ft. 9 in. Top speed is 466 mph with a loaded weight of 135,584 lb. The newer IL-18e version has more powerful turbines. It appears to have been ordered by the Red Chinese. In spite of the political split between these two Communist nations, trade still prospers. Newer models of the IL-18 may be included, as well as other civil transport types, in a substantial Sino-Soviet trade agreement signed in Moscow on July 27, 1967.

Ilyushin IL-28 Medium Bomber

For almost twenty years the standard medium bomber and attack plane of Communist air forces throughout the world, the Ilyushin IL-28 was the first mass-produced post–World War II jet bomber of the Soviet Air Force. First flown in prototype form in 1947, powered by two British Rolls-Royce Nene turbojets, the IL-28 began to enter Russian service in numbers in 1949. The IL-28 was a very straightforward design. Its later models mounted two Klimov VK-1 turbojets of 5,950 lb. thrust in slab-sided nacelles on a straight wing spanning 68 ft. 2 in. Only the tail surfaces are swept in modern jet fashion, giving the IL-28 a rakish appearance. The IL-28 can carry a bomb load of 4,400 lb. at a speed of 580 mph, certainly not fast for these days of supersonic flight but sufficient to worry any target defense. A crew of three or four is carried, and the tail gunner sits in splendid isolation in a compartment so large it makes the aircraft look tail-heavy. The tail is protected by two manually operated 23 mm cannon while two 20 mm or 23 mm cannon are mounted in fixed position in the nose. The IL-28 was first introduced to the West in dazzling fashion when more than fifty of them flew over Red Square in Moscow on May 1, 1950. Communist China was the largest export customer for the IL-28; the first examples arrived late in 1952 to pose a distinct threat to the U.N. forces in South Korea. The Red Chinese never used this force over South Korea, in fear of massive American retaliation. As a result the PLAAF jet bomber regiments survived the Korean War, only to pose a continued threat to the American Seventh Fleet off Taiwan. Russian deliveries of the IL-28 continued throughout the 1950's, and approximately 160 were still estimated to be in Chinese service in the late 1960's. Ten of these aircraft were delivered to Pakistan by the Chinese in 1966. NATO first gave the IL-28 the code name "Butcher," then later changed it to "Beagle," an apt description of the IL-28's appearance.

Ilyushin IL-28U Medium Bomber Trainer

The best way to ruin the attractive appearance of an IL-28 "Beagle" bomber would be to stick an extra cockpit right on its smooth nose, and that's just what the Russians did to create the IL-28U conversion trainer model for the training of pilots, bombardiers, and crew members. First appearing in 1951, the

IL-28U was virtually identical in size and weight to the bomber version except for the extra cockpit and the removal of the fixed nose armament. It was supplied in small numbers to the nations flying the IL-28—the People's Republic of China, Czechoslovakia, Rumania, Poland, Afghanistan, Indonesia, and the United Arab Republic, among others. The nose of the trainer version extended beyond the additional cockpit to create the only basic dimensional change. The training cockpit is fully equipped with pilot's controls and sits below and in front of the true pilot's cockpit, giving the instructor a good view of the student pilot while in flight. While essentially a training machine, examples of the IL-28U are posted to the PLAAF air divisions flying the IL-28 bomber so that these units can continually check on the proficiency of their pilots and crews. The NATO code name is "Mascot."

Junkers-Fili F.13 Transport

One of the stipulations of the Treaty of Versailles following World War I, signed on June 28, 1919, was that Germany could no longer produce or maintain military aircraft. German aircraft manufacturers, not wishing to go out of business in what promised to be one of the most successful ventures of the post–World War I world, set up manufacturing plants in neighboring countries. As a result, Sweden, Switzerland, and the Soviet Union found themselves with complete aircraft-producing facilities staffed by Germans and employing both German and local labor. The Junkers plant in Russia, located at Fili just outside of Moscow, was the most modern aircraft factory in the Soviet Union. It produced Junkers transports, bombers, fighters, and trainers. The world-famous F.13 monoplane design of 1919, the first all-metal low-wing transport to be built, was among the original aircraft produced at the Russian plant. It was powered by a Russian-built Junkers L-5 water-cooled engine of about 280 hp. The F.13 carried a crew of two and as many as six passengers, or an equivalent amount of freight. It became the standard Russian military transport and was also exported to other war lords in China in addition to the combined Communist and Kuomintang forces under the command of Chiang Kai-shek in the middle 1920's.

Kawasaki Ki.48 Type 99 Two-Engine Light Bomber

Standard light bomber of the Japanese Army in World War II, first entering production in 1939, the Ki.48 served in China throughout the war. Together with the Mitsubishi Ki.51 single-engine attack plane, it handled the bulk of the bombing duties in north China and was therefore available in numbers after the war ended. Examples were picked up by both the Chinese Nationalists and the Chinese Communists, and it flew in both of their air forces. Late production models of the Ki.48 were fitted with wing brakes for dive-bombing operations, but the Japanese Army crews usually removed them to avoid the maintenance and boost the top speed slightly. It wasn't a very large aircraft, having a span of just over 57 ft., yet somehow it looked much larger. Fully loaded, it weighed 14,500 lb., including a 1,600 lb. bomb load. Nearly 2,000 of these somewhat ordinary bombers were built. In its Model IIb version the top speed was 308 mph provided by two Nakajima Ha.115 14-cylinder radial engines of 1150 hp each. The design format was that of a mid-wing monoplane with a normal crew of four. Armament was two or three 7.7 mm machine guns with one in the nose and either two or three in dorsal positions at the back of the cockpit hood. United Nations wartime code name for the Ki.48 was "Lily."

Kawasaki Ki.61 Type 3 Hien Fighter

The favorite fighter of many Japanese Army pilots, and preferred by many of them to the Ki.43 built by Nakajima, the

Kawasaki Ki.61 was the only World War II Japanese fighter to enter production and combat service utilizing a liquid-cooled inline engine. With its low wing, pointed nose, and radiator under the fuselage, the Ki.61 was easily mistaken for an American P-51B Mustang, confusing both the Japanese and the Americans during the Pacific War. The Japanese called this fighter the Hien, or Swallow, and its wartime United Nations code name was "Tony." It was fast; its early models with the 1100 hp Kawasaki Ha.40 engine reached a speed of 350 mph. This powerful inline engine was a Japanese development of the German Daimler-Benz 601A, a type that was licensed for production in Japan. The Ki.61 Hien followed the Japanese Army Air Force wherever it was stationed, and "Tony" fighters were sighted over Japan, China, Southeast Asia, and the Pacific outposts. Quite a number of them were stationed in China, and both the Nationalists and the Chinese Communists picked up examples in good condition. The Ki.61 had a wing span of 39 ft. 4¼ in. Over 3,000 of these fighters were produced in Japan between 1943 and 1945.

Lavochkin LA-9 Fighter

Final World War II development of a long series of clean-looking Russian wartime fighters, the LA-9 was an all-metal version of the previous Lavochkin LA-7 model, a fighter type with a wooden airframe. As such, the LA-9 demonstrated the growing sophistication of the Russian aircraft industry. While the wartime Axis powers of Japan and Germany desperately tried to produce their most advanced designs in wood to overcome their shortages of aircraft dural and other metals late in the war, the Russians were doing just the opposite, having used wood for many years. This shift in materials usage, with the concentration on metals, was to have a profound effect on Russian airpower in the immediate post–World War II years, for it upgraded the Russian aircraft productive capability just when the age of jet aircraft was beginning. The LA-9 barely made it into action in the Great Patriotic War, or GPW, as the Russians called World War II, and when the war ended in August, 1945, it was being delivered as the new standard long-range fighter of the Soviet Air Force. In the same class as the American North American P-51D Mustang, the LA-9 mounted a 14-cylinder Shvetsov ASh-82FNV radial engine of 1850 hp, giving it a top speed of about 400 mph. The LA-9 had a wing span of 34 ft. 9¼ in. It was among the first aircraft supplied to Communist China after the end of the civil war in 1949 and remained in Soviet and Communist Chinese service for many years and ultimately received the NATO code name "Fritz."

Lavochkin LA-11 Fighter

A further refinement of the LA-9, making its debut in Russia in 1946, the LA-11, which equipped Russian squadrons in the early years of the Cold War, became the last piston-engine fighter of the Soviet Air Force. Since it was the last example of reciprocating-engine fighter development in the Soviet Union, Western experts were overjoyed to get an opportunity to study one that a defecting Soviet pilot crash-landed in Sweden on May 17, 1949. They were reportedly shocked at the crude workmanship on the LA-11, and concluded that the Russian aircraft industry was years behind that of America and Great Britain, leading to an almost paranoid surprise when modern Russian jet fighters showed up in quantity over Korea only a little more than a year later. The reportedly crude workmanship of Soviet aircraft seems to be a permanent Russian trademark, for even the modern MiG-21 fighters of the present day seem to present the same picture to Western technical experts. This crude workmanship doesn't seem to have hurt Russian aircraft performance over the years and may well be a Russian production advantage rather than a fault. The LA-11 fighter used the same Shvetsov ASh-82FNV power plant as the LA-9, but weighed over 500 lb. less because of careful redesign work. Armament of the LA-11 was also lighter, the newer fighter carrying three 20 mm cannon rather than the four cannon of the LA-9. The result was an increase of over 20 mph in top speed in a tougher, more maneuverable fighter. First delivered to Red China in 1950, the LA-11 became one of the standard PLAAF fighters just prior to the Korean War and served as a second-line PLAAF fighter well into the 1950's. NATO code name for the LA-11 is "Fang."

Lavochkin LA-15 Fighter

A contemporary of the Mikoyan MiG-15 and developed in parallel with the MiG design, the LA-15 was created by Russian designer Semyon A. Lavochkin to utilize the Rolls Royce Derwent centrifugal-flow turbojet of approximately 3,500 lb. thrust, a power plant that had been obtained from Great Britain in 1947 in a sale that created an international stir. Looking very much like a MiG-15 with its wing moved up to the shoulder of the fuselage, the LA-15 was intended as a back-up fighter design in the event that the more powerful MiG-15 design didn't work out. Success with the prototypes in 1948 led to production orders, and the LA-15 entered Russian service in some numbers. It first appeared to Western observers when a flight of three flew overhead during the annual Tushino display in August, 1949. The LA-15 was slower than the MiG-15 and was soon passed over in favor of further development of the MiG line. An attempt was made to soup up the LA-15 design by mounting the more powerful Klimov RD-45 jet engine used in the MiG-15 although this model, given the LA-168 design designation, was not selected for production. The LA-15 soon disappeared from Russian service as the remaining examples were exported, a few going to Communist China for use in Korea. An example remains on static display in Russia on the grounds of the Soviet Air Force Academy at Monino.

Lisunov LI-2 Transport

As the Soviet Union developed its own indigenous aircraft design capability in the 1930's and concentrated on the production of aircraft types developed in its own design bureaus, the habit of producing foreign models was slowly but surely discontinued. But the Russians did keep their eyes open for the latest advancements in the field of aviation, and once in a while paid hard cash for the opportunity to build modern foreign types in the Soviet Union. One of these notable exceptions in the mid-1930's was the Soviet selection of the American twin-engine low-wing monoplane Douglas DC-3 transport for production in Russia. The basic design of the DC-3 had so completely revolutionized civil air transportation throughout the world by its durability, speed, and carrying capacity that it virtually became the world's leading transport. Douglas sold complete production and technical-aid licenses to the Soviet Union as well as to Japan, and supplied the necessary drawings, jigs, machine tools, and resident engineers to get production started efficiently in both nations. The Soviet Union and Japan both quickly saw the military airlift capabilities of the DC-3, and independently developed military cargo and passenger transports similar to the later American C-47. In Russia the conversion design work was undertaken by an engineer named Lisunov, and the military transport model entered production as the PS-84 to become the standard Russian military transport of World War II. The Russians put their cargo door on the right side and added a power turret to the top of the fuselage in some models. Otherwise it was virtually identical to its American C-47 counterpart. Late in World War II the type was redesignated the LI-2 at a time when most Russian aircraft were renamed to give their designers the proper credit. Russian LI-2 transports were supplied to Red China in substantial numbers prior to and during the Korean War to build up the PLAAF airlift capability. Civil models are also in use with CAAC, the Red Chinese national airline. Power for the Lisunov LI-2 was supplied by two Russian M-63 radials. The wing span was 95 ft. The LI-2 has received the NATO code name "Cab."

Loening-Keystone Amphibian

One of the geniuses of American aviation at the end of World War I and throughout the 1920's, Grover Loening originated hundreds of the basic ideas that contributed to the modern aircraft of the 1930's and 1940's. Loening aircraft were known for their original design features and their reliability. The Loening eight-passenger Amphibian brought high performance to military and commercial seaplanes for the first time and put them on a cost-comparison par with landplanes. China was a big Loening-Keystone Amphibian customer; eight Amphibians being purchased between 1929 and 1933 by China National Aviation Corporation, the first economically viable civil passenger, freight, and mail airline in China. Power for the Chinese examples was either a 525 hp Pratt and Whitney Hornet or a Wright Cyclone. The sale is unrecorded, but a number of Loening-Keystone Amphibians were used by the Nineteenth Route Army in Fukien for bombing and reconnaissance missions in 1933. In appearance the Loening-Keystone Amphibian was a biplane that looked very much like a landplane type, except for its long single float reaching forward from the nose of the aircraft. The cockpit was open, but the passengers were carried in a cramped but not uncomfortable cabin. Many influential Chinese made their first trip into the air in a Loening-Keystone, including the Dali Lama of the early 1930's.

Manshu Ki.79 Type 2 Advance Trainer

Companion trainer to the Tachikawa Ki.55 of the Japanese Army Air Force, and built to serve the same function as a standard Japanese Army advance trainer, the Manshu Ki.79 utilized the airframe of an earlier Japanese fighter. The parent aircraft was the Nakajima Ki.27 Fighter, a single-seat low-wing monoplane with a fixed landing gear that entered production in 1937, and the first Japanese army fighter to have an enclosed cockpit cover. In January, 1942, after the Ki.27 fighter had passed its zenith and was replaced on the Nakajima production lines by the Ki.43, the Japanese Army selected the basic airframe for conversion to an advanced trainer. The simple lines of the Ki.27, the fixed landing gear, and the availability of production jigs for a trainer type much needed by the rapidly expanding wartime Army Air Forces of Japan led to its acceptance as a two-seat and single-seat advance trainer type in 1942. The aircraft was modified with lower-powered engines than the earlier fighter and with dual controls in its two-seat versions. Production of the new trainer was assigned to the Manshu Airplane Manufacturing Company in Manchukuo, a complete aircraft manufacturing facility set up in the Japanese satellite nation with Japanese funds. One- and two-seat open cockpit versions were produced in a profusion of four different models, with examples of all models later picked up by the Chinese Communists at Mukden. Power plants used were various models of the Hitachi Ha.13A 9-cylinder radial ranging from 470 hp to 515 hp. The single-seat Ki.79a Model A version could hit a top speed of 225 mph. Manshu built over 3,700 of these trainers.

Mikoyan MiG-9 Fighter

When the MiG-9 first caught the eye of British and American observers as it flew over Red Square in Moscow on May

1, 1947, it was thought to be a research aircraft designed by the Lavochkin Design Bureau. It was not yet two years since the end of World War II, and the thought that the Russians had developed jet aircraft beyond a few test airframes seemed an impossibility. But the MiG-9 was a far more significant aircraft; the examples that took part in the military air display on May 1 were actually service test examples of the Soviet Union's first mass-produced fighter created from the ground up as a jet aircraft. It was also the first of the fabulously successful MiG jet series, the products of the design team of Artem I. Mikoyan and Mikhail I. Gurevich, whose combined initials gave the MiG fighters their short, staccato name. Many of the later MiG fighter features that were to become so famous over the years were first contained in the MiG-9: the mid-wing layout, short nose, nose inlet, high tail section and distinctive MiG cockpit hood. The power plants for the MiG-9 were two RD-20 axial-flow turbojets just short of 2,000 lb. of thrust each, Russian developments of German Junkers and BMW jet engines captured in Germany at the end of the war in Europe. Top speed of the MiG-9 was nearly 600 mph. Four-hundred-seventy of these early jet fighters were built, and some were reportedly delivered to Communist China to become the first jet aircraft of the PLAAF. Although their service life was short, they paved the way for the more advanced MiG jets that were to show up a few years later. The NATO code name for the MiG-9 is "Fargo."

Mikoyan MiG-15 Fighter

If a vote were to be taken for the most famous jet fighter aircraft in the world, the mid-wing Russian Mikoyan MiG-15 would probably win. From its introduction to the world at large in the hands of Red Chinese pilots during the Korean War in 1950 and 1951 to the present day its designation has become synonymous with Russian influence, intransigence, and aggression. In its middle-1950's heyday it was one of the two or three best jet fighter aircraft in the world. It was built in greater numbers than any jet fighter before it—production was estimated at more than 17,000 units. The original production models were powered by a 4,960-lb-thrust Klimov RD-45, a Russian adaptation of the British Rolls Royce Nene centrifugal-flow turbojet, a number of which were sold to the Soviet Union in 1947 along with examples of the Rolls Royce Derwent, a jet engine of similar design but lower thrust. The sale of these jet engines to the Soviet Union at a time when the Russians were tightening their grip on Eastern Europe led to some bitterness in the West. It is probably true that the sale saved the Russians months of development time, but the Soviet Union would inevitably have created a jet air force in any event to keep up with Stalin's postwar goals in Europe and Asia. The MiG-15 was an aircraft perfectly suited to Stalin's ambitions. Its controlled speed of Mach 0.91 was faster above 30,000 feet than the American F-86E, its major adversary in history's first jet warfare over Korea. It could also climb faster and had a higher ceiling. Smaller than most fighters, its span was only 33 ft. ¾ in. Originally coded "Falcon" by NATO, its code name was later changed to "Fagot."

Mikoyan MiG-15bis Fighter

The outstanding success of the MiG-15 fighter quickly led to further development and increased power to take full advantage of a winning design. The key step in creating the MiG-15bis (with the identification "bis" borrowed from the Italians and French to mean second, or improved, model) was the further development of the Klimov RD-45 centrifugal-flow jet engine to increase its thrust by 20 per cent to 5,950 lb. with a boost to 6,750 lb. with water injection. Fuel capacity of the MiG-15bis was increased by adding internal tanks as well as a pair of wing-mounted drop tanks. Additional electronic gear was also added; yet the total weight was cut by about 200 lb. as a result of design improvements throughout the airframe and a drastic reduction in the number of individual parts required for assembly. The MiG-15bis retained the armament of two 23 mm cannon and a single 37 mm cannon mounted under the nose of the late production models of the earlier MiG-15. The MiG-15bis made its combat debut over Korea in Red Chinese hands in 1952, following General Liu Ya-lou's wartime trip to Moscow for further Russian aid to the Chinese Air Force. Quite a number are in service throughout the world to this day. The wing span of the MiG-15bis is 33 ft. 1½ in., and its loaded weight is 11,085 lb. Its rated top speed of 668 mph made it one of the fastest fighters in its prime. In its role as a fighter-bomber, currently its most widespread activity, it can be loaded to 13,889 lb., although performance drops drastically. The large inventory of MiG-15bis fighters in Communist China, Albania, Cambodia, and North Viet-Nam has created a continued demand for spares and maintenance parts, a situation met admirably by the Chinese Communists. A special department of the National Aircraft Factory at Shenyang and subcontractors throughout mainland China, still produce parts for this ageless fighter. The NATO code name, as for the MiG-15bis "Fagot."

Mikoyan MiG-15UTI Fighter Trainer

The jump from a piston-engine trainer to a jet fighter is a long step, and the Russians soon found, as did most advanced aircraft-producing nations, that the transition could be eased by the creation of a two-seat trainer version of the actual fighter aircraft. The MiG-15bis was picked as the likely subject and the MiG-15UTI trainer soon became standard equipment in virtually every air force that flew, or still flies, the various models of the MiG-15 and MiG-17 fighter series. Designated the MiG-15UTI, the trainer MiG has an awkward hunchbacked look created by its oversized canopy designed to fit over both pilot instructor and student. This feature alone cost the MiG-15UTI enough speed to make it considerably slower than its MiG-15bis fighter counterpart. Power is reportedly provided by an improved Klimov RD-45 of 5,450 pounds thrust, and the span and length are identical to the MiG-15bis. Although the MiG-15UTI is a trainer, an armament of two 23 mm cannon is retained on some examples in case of military duty. The MiG-15UTI received immediate acceptance in the Communist air arms to which it was supplied, and licenses for its production were allotted to Poland, Czechoslovakia, and the People's Republic of China. Many of these trainers are still in service and no doubt will be as long as the MiG-15bis and the MiG-17 remain in use as fighters. Deliveries of the MiG-15UTI were still being made by the Russians in the middle 1960's, although the more advanced MiG-21UTI trainer version of the advanced MiG-21 series is replacing the earlier trainer in many of the air forces receiving Russian equipment. NATO gave the MiG-15UTI the code name "Midget."

Mikoyan MiG-17 Fighter

While the MiG-15 and MiG-15bis fighters were making their mark in the skies over Korea, development work proceeded in the Soviet Union on a completely re-engineered version of the basic MiG fighter design as the MiG-17 (a situation reminiscent of the re-design of the earlier Yakovlev YAK-15 fighter as the YAK-17). A fresh approach was taken with the MiG-17 wing structure and the airframe, although the successful formula of the parent MiG-15 series was retained. The powerful 5,950-pound-thrust Klimov RD-45/VK-1 engine of the MiG-15bis was retained, with afterburning becoming standard in the later MiG-17 models. Compared to the wing of the MiG-15bis, the MiG-17 wing was thinner, had a shorter span (31 ft. 6 in.), and was swept back farther. The fuselage was refined and lengthened below the vertical tail, although the measured aircraft length was an inch shorter than the earlier MiG. The initial Day Fighter models of the MiG-17, coded "Fresco-A" by NATO, showed up in the Soviet Air Force in 1952. Armament was the same as the MiG-15bis, consisting of two 23 mm cannon and a 37 mm cannon, all in the nose of the under fuselage. An improved "Fresco-B" version next appeared with revised air brakes, and was followed by the Klimov VK-1A–powered MiG-17f "Fresco-C" with a turbojet rated at 6,990 lb. thrust with the afterburner in operation. A radar-equipped version of the MiG-17f ("f" meaning "forsazh," or Russian for boosted–relating to the increased power) showed up in 1955 for all-weather interception and received the NATO code name "Fresco-D." The models supplied to Red China were primarily "Fresco-A" and "Fresco-C" versions, although Chinese Communist reports of dirty-weather interceptions suggest that a fair number of MiG-17f "Fresco-D" models may also have been delivered. Flown by more air forces around the world than any other Russian fighter type and produced under license in Poland, Czechoslovakia, and Red China, the MiG-17 and its variants have been built in greater numbers than any other jet fighter, a record that may stand for all time now that fewer aircraft are needed in war to achieve their objectives.

Mikoyan MiG-19 Fighter

Russia's first supersonic service fighter, and an advanced development of the MiG series, the Mikoyan MiG-19 differed from the earlier design format in having two turbojet powerplants. Power was supplied by two Klimov RD-9F/VK-7 axial-flow turbojets of 6,700 lb. thrust each, giving the MiG-

19 almost twice the power of the earlier MiG-17. In typical Russian fashion, the MiG-19 was introduced to the West in a dramatic flyby at Tushino in 1955 when forty-eight of the fighters flew over the viewing stands in formation. The point was forcefully made that the new fighter was entering Russian service in quantity. Although the MiG-19 was obviously an advanced model in the MiG-15bis and MiG-17 series, it was far more than a modification. The higher supersonic speeds required a completely new structure, and so the MiG-19 was created as a new design from the ground up. The prototype flew in 1953, but it was five years before the type was considered for service in Communist China. The Russians typically export their "latest" fighter only when a new follow-on fighter is available for exclusive Soviet use, and the coming of the MiG-21 eventually released the MiG-19 for export. The MiG-19 never became as well known as the earlier MiG fighters, or as the later MiG-21, in fact. At best it was a transitional type, pushing the state of the art forward in the middle 1950's, but it was soon replaced by the faster, more powerful fighters of the present era. The MiG-19 has a single-seat mid-wing form, and introduced the radar gunsight to Russian, and ultimately to Communist Chinese, service. The MiG-19 entered service in the Czech, East German, and Polish air forces in the late 1950's in addition to Red China, and later showed up in Cuba, Mongolia, Egypt, Iraq, Syria, and Indonesia. Later models were fitted with Klimov VK-9 turbojets of 7,850 lb. thrust each, which further boosted power and pushed the top speed just over 900 mph. Wing span is 29 ft. 6 in. and length 41 ft. 4¾ in., so that the MiG-19 has a long, lean look. Empty weight is 12,132 lb. and the fighter can be overloaded to a maximum of 22,500 lb. Normal armament is two 23 mm cannon. Most of the MiG-19 fighters in the world have been relegated to low-level attack and fighter-bomber duties, although the Red Chinese still use them as first-line interceptors. They have been spotted along the Red Chinese

perimeter just north of the North Vietnamese border, waiting for American "strays" to cross the border and feel the wrath of the PLAAF. NATO code name for the MiG-19 is "Farmer."

Mikoyan MiG-21 Day Fighter

The MiG name seems to have an extraordinary fascination for the world. Perhaps its brevity makes it appealing—or the fact that it sounds strong. In any event the best known combat aircraft of the past two decades have both been MiGs. During the Korean War and the later 1950's the designation MiG-15 was a household word in both the Communist and the non-Communist worlds. No single military aircraft threatened to earn this distinction until the MiG-21 came along in the 1960's, and already it seems destined to become as well known as its earlier namesake. Originally conceived in the Soviet Union right after the Korean War as a replacement for the then leading MiG-15 and MiG-17 fighter series, the MiG-21 was designed as an interceptor for the defense of Russian cities and industrial targets. Two test airframes were produced in 1955 differing primarily in their wing structures, one having thin swept wings and the other a delta wing format. Both airframes had conventional swept-tail surfaces. When the two prototypes made their appearance over Tushino on June 24, 1956, observers recorded them as distinct types, giving them the NATO code names "Faceplate" and "Fishbed." Although both versions were shown at the Moscow air show, the Russians had already selected the delta "Fishbed" version for further development, and flight testing of improved military prototypes began in 1957. The Russians quite probably displayed the prototype MiG-21 proudly to Mao Tse-tung when he visited Moscow for the World Communist Conference in November, 1957, a move which would certainly have led to Red Chinese interest in the aircraft for the PLAAF.

Smaller and lighter than its American counterparts, the "Fishbed-B," as the first production version was coded by NATO, had a span of about 25 ft., a length of some 40 ft., and a loaded weight of approximately 17,000 lb. The "Fishbed-B" began to enter Soviet service early in 1959, and within a year examples were secretly delivered to the Communist Chinese and plans were initiated to produce the aircraft in China, mute testimony to the close relationship between these two Communist powers before their split. Although the exact delivery dates are unknown, it is highly possible that the People's Republic of China was the first Communist nation outside of the Soviet Union to receive what was then Russia's latest fighter. For some time after the first MiG-21 fighters were sent to Red China the Chinese Communists specifically requested quantity deliveries of the new fighters in order to re-equip the PLAAF. But by then the tide of events had turned. The Russians not only ignored Peking's requests but compounded the insult from the Red Chinese point of view by exporting later models of the MiG-21 to more than a dozen Communist and non-Communist nations, including the surrounding Asiatic states of North Korea, India, and North Viet-Nam. In its "Fishbed-B" production version the MiG-21 was reportedly powered by a RD-9 axial-flow turbojet of about 9,500 lb. thrust. The top speed of this model was in the neighborhood of Mach 1.7 or about 1,120 mph at 40,000 feet.

Mikoyan MiG-21f Day Fighter

The most widely used version of the MiG-21 series outside the Soviet Union in the late 1960's, and one of the leading fighters flown by the small air arm of the Democratic Republic of Viet-Nam, the MiG-21f Day Fighter is perhaps the best known model in the "Fishbed" series. The "f" designation does not mean that it is the sixth aircraft in the line, as it normally would by American or British standards. The Russians are far more pragmatic in their aircraft designations, and the letter "f" stands for "forsazh," or "boosted," indicating that this second service version of the MiG-21 has greater power than the original model. Model improvements within the series are then handled by digits following the designation. The Russian practice of describing basic airframe changes in their designations has caused some confusion in the identification of the aircraft. The NATO code name for the MiG-21f is the "Fishbed-C," which has often caused the incorrect designation "MiG-21C" in the Western press. This wouldn't be so bad by itself, but the Russians continue to improve the type, and already NATO-coded "Fishbed-F" and "Fishbed-G" models are in service. The result is the confusion of having two completely different MiG-21F fighters. The MiG-21f (Russian style) is in use around the world. It was one of the prime fighter types used by Egypt, Syria, and Iraq in the three days of air fighting in the Middle East war of June, 1967. At that time the MiG-21f was largely discredited at the hands of the pilots of the Israeli Air Force flying French Dassault Mirage IIICJ fighters. This was actually unfair, for even the Israelis reported that the MiG-21C (American style) was an efficient high-altitude fighter. They should know, for a defecting Iraqi pilot landed his MiG-21f-13 in Israel the previous August and thus gave the Israeli Air Force the opportunity to fly the aircraft in mock combat against their own pilots in order to train them for the coming war. In spite of their praises for the MiG-21C, the Israeli pilots felt it was easier to shoot down than the more maneuverable Mikoyan MiG-17, an earlier MiG type also flown by U.A.R. pilots. As one of the principal antagonists facing American pilots over North Viet-Nam, and a possible opponent in Red Chinese hands as a result of their pilferage of North Vietnamese examples, the MiG-21f is of particular importance to the American Air Force and Navy. Its secrets have finally been revealed. In July, 1967, three examples captured by the Israelis were picked up by USAF transport planes. Two of them are being flight tested at Edwards Air Force Base while the third is under dissection at Wright-Patterson Air Force Base. The MiG-21f is reportedly powered by a 13,000 lb.-thrust R-37f axial-flow turbojet with an afterburner, giving the aircraft a Mach 2.1 speed full-out. Although the wing did not seem to change from the earlier "Fishbed-B" version, the fuselage was modified and extended to about 47 ft. The "Fishbed-C" first appeared in Russian service in 1961, and soon showed up in the Eastern European Communist satellite nations as well as in India, Indonesia, the Democratic People's Republic of Korea, Finland, Cuba, and other nations until it is now numerically one of the world's most important fighter aircraft. The type is also being built under license in Czechoslovakia and India. The export versions are delivered with manuals written in English, and the same language is used to identify various instruments and provide instructions on the actual aircraft. Taking a MiG-21f up for a flight is not quite as easy as it would appear, because the Russian translations into English are somewhat confusing.

Mikoyan MiG-21Pf All-Weather Fighter

A fast look at the MiG-21Pf suggests that it is a MiG-21f with a pointed nose coming out of the air intake, but a closer look reveals that this all-weather fighter-interceptor model is considerably different from the earlier version. The fuselage has been completely redesigned to accommodate the required electronic gear, and the nose radome extends the length of the aircraft to about 49 ft. The power plant appears to be the same 13,000 lb.-thrust R-37f used in the MiG-21f, because of the retention of the "f" designation, although the refinements in design have reportedly pushed the MiG-21Pf up to Mach 2.2 in speed. A substantial number of these dirty-weather fighters have been supplied to the Democratic Republic of

Viet-Nam, and the Russians have reported that some of them were expropriated by the Chinese from the supply line stretching from the Soviet Union to North Viet-Nam through Communist China. The MiG-21Pf is also in service in East Germany, Czechoslovakia, Poland, Egypt, Syria, and Indonesia, and will probably reach service in other Communist and non-Communist nations around the world. A number of them were destroyed by the Israeli Air Force on the ground in the short Middle East war of 1967. The cannon arrangement of the earlier MiG-21 fighters has been dropped in the MiG-21Pf, and infra-red homing missiles have been mounted on external pylons on the wings. American experience with the "Fishbed-D" (NATO code name for the MiG-21Pf) in air fighting over North Viet-Nam indicates that the fighter is not able to take much punishment and is outclassed by the F-4 Phantom. Red Chinese acquisition of the type suggests that an all-weather version of the Chinese MiG-21 may soon make its appearance, providing a logical and necessary addition to the PLAAF's defense capability. Quite obviously, the MiG-21Pf is far from the end of the line for this versatile series of fighter aircraft. A further model with a refined fuselage and wider-chord vertical tail surfaces was in wide-scale Russian use early in 1967, and also formed the equipment for the two best-known Soviet Air Force aerobatic teams. This version received the NATO code name "Fishbed-F." The MiG-21 series may even have a new lease on life, for an STOL (short-take-off-and-landing) version with direct-lift turbine power centrally mounted in the fuselage was demonstrated at the Soviet "Aviation Day" Air Show just outside of Moscow at Domodedovo on July 9, 1967. Perhaps the Chinese Communists will also pick up this lead, and thus have an additional model to innovate with their Shenyang version of the MiG-21 series.

Mil MI-1 Helicopter

If you didn't know that the Mil MI-1 was a Russian helicopter design, you could easily mistake it for an early American Sikorsky design, or a blood brother of the Sikorsky S-51. Aircraft technology in the years immediately following World War II was at such an even keel—designers in the victorious nations worked on the same track around the world, and copies of captured Japanese and German aeronautical documents were available to all—that new designs of the late 1940's and early 1950's looked very much alike, no matter which nation produced them. Russian, American, and British fighters of the period could almost have been cut from the same mold. But nowhere did this similarity show up as keenly as it did in the field of helicopters. The Mil Helicopter Design Bureau was established in 1946 under the direction of Mikhail L. Mil, sometimes called the Sikorsky of Russia, after Igor I. Sikorsky, the American helicopter pioneer, himself a Russian by birth. Created for a wide variety of roles as a utility helicopter, the MI-1 became the Soviet Union's first mass-produced helicopter. The MI-1 was first seen in numbers at the Russian Aviation Day show at Tushino Airport outside of Moscow on July 8, 1951. It became Russia's most widely produced military and civil helicopter and was exported to virtually all Communist nations as well as sold on the open market to Indonesia, the Arab nations, and other non-Communist buyers. Advanced models were still in production in Russia and Poland in the late 1960's. The basic design was given a new start in the middle 1960's with the development of a MI-2 model in Russia powered by two turbine engines. The Mil MI-1 models sold to Red China had a loaded weight of 4,938 lb. and were powered by 575 hp Ivchenko AI-26GRF engines. NATO code name for this earlier model is "Hare."

Mil MI-4 Helicopter

The most widely used helicopter in Russian service, and throughout the Communist world, the Mil MI-4 has been in production since 1952. In layout it is an enlarged development of the earlier MI-1 design, mounting a four-bladed main rotor 68 ft. 10¾ in. in diameter and a three-bladed tail rotor driven by a 1,700 hp Shvetsov ASh-82V 14-cylinder radial. The MI-4 can seat fourteen passengers or carry a cargo of 2,645 lb. at a top speed of 130 mph. The Russians quickly realized that this well-built aircraft had great export potential, and aggressive sales efforts have been made with some degree of success. Substantial numbers went to Communist China after the Korean War. The first sale to a Western nation was to Austria for civil use in 1959, with deliveries also being made to India, Egypt, Finland, and a number of other non-Communist nations. This sales campaign is still going on with the Russians offering liberal terms and low prices, and it looks as if the MI-4, as well as the later Russian MI-6 and MI-10 helicopters, will be flying throughout the world for many years to come. It would appear that the real penetration of the western markets is just beginning. The MI-4 has an impressive record behind it. In 1956 it set three important load-carrying records for helicopters, and many of these Russian helicopters have served well in climates ranging from the tropics in Viet-Nam to the sub-zero temperatures of the Arctic, where MI-4 helicopters are used by Soviet scientific missions. The MI-4 looks so much like an American Sikorsky S-58 that the two types are often mistaken for each other. The NATO code name for the MI-4 is "Hound."

Mitsubishi Ki.51 Type 99 Ground Attack

Japanese involvement on the Chinese mainland, and the continuous state of guerrilla warfare faced by their ground forces there between late 1937 and August, 1945, led to the requirement for counter-insurgency aircraft for use against small mobile Chinese resistance units. A virtually identical situation was faced by the United States in Viet-Nam over twenty years later. The somewhat old fashioned looking Mitsubishi Ki.51 Type 99 ground attack and reconnaissance aircraft of 1939 fit the bill perfectly for the Japanese, and did the same type of job that the ageless North American T-28D Trojans and Douglas A-1 Skyraiders did in Viet-Nam in the 1960's. Mitsubishi was one of the three major aircraft producers in Japan during the Pacific War, as the Japanese called the extension of their war with China into the Pacific after Pearl Harbor. Although the firm was noted for its modern fighters and bombers, not the least of which was the famous wartime "Japanese Zero," the Mitsubishi series of fixed-landing-gear aircraft remained in production during the war years to meet the needs in China. The Ki.51 was an advanced development of the earlier Mitsubishi Ki.30, although it was somewhat more streamlined and had a more powerful engine. Use of the fixed landing gear on the Ki.51 may have seemed archaic, but it had a very practical purpose: it kept the aircraft relatively simple to maintain at fields deep in China. With its crew of two and a bomb load of 440 lb. of small anti-personnel bombs, the Ki.51 was able to reach a top speed of 264 mph. Power plant was a Ha.26-II 9-cylinder radial of 900 hp. Examples captured by the Chinese Communists in Manchuria and North China served as tactical bombers in support of the PLA during the civil war from 1946 to 1949. Wartime U.N. code name was "Sonia."

Mitsubishi Ki.57 Type 100 Transport

A clean-looking twin-engine low-wing monoplane transport of Japanese design, the Mitsubishi Ki.57 was the standard Japanese Army Air Force transport. It served with JAAF transport units and with virtually every Japanese Army air combat regiment of World War II. Utilizing the wings and tail surfaces of the then standard Mitsubishi Ki.21 Heavy Bomber, with a fuselage that followed the successful Douglas DC-3 transport format, the Ki.57 entered production in Japan in 1940. It was the top choice of Japanese Army generals for their personal VIP transport because it was fast, with a top speed of 297 mph, and large enough for its interior to be luxuriously fitted out with a field office, dining area, and sleeping quarters. Its span of 74 ft. 1½ in. made the Ki.57 considerably smaller than its American Douglas C-47 counterpart. The Ki.57 was also widely used as a civil transport in its passenger Mitsubishi MC.20 version by Japan Air Transport, the Japanese National airline; Manchukuoan Aviation Company in Manchukuo; and by China Airways Company, the flag airline of the National Government of China, the wartime Japanese-controlled puppet government of occupied China with its capital at Nanking. A substantial number of both the military and the civil versions were picked up by the Chinese Communists and the Nationalists. During World War II the Ki.57 received the U.N. code name "Topsy."

Nakajima Ki.34 Type 97 Transport

Built in prototype form in 1936, the Ki.34 was the first original Japanese low-wing monoplane transport with a retractable landing gear. The design was actually a ¾-scale version of the American Douglas DC-2, the forerunner of the later DC-3, and widely used on American airlines in the mid-1930's. The Douglas DC-2 was also produced in Japan by Nakajima under a Douglas license, providing the Nakajima firm with the pattern for their own original transport. The 65 ft. 4 in. wing span of the Ki.34 made it considerably smaller than the American transport DC-2, which had a span of 95 ft. The Ki.34 entered production in 1937, and a total of 299 of them were built for Japanese Army use as a passenger and paratroop transport between 1937 and 1942. Much of the production was handled by Tachikawa in order to free the Nakajima facilities for the production of combat aircraft during the war years. The Ki.34 carried a crew of two or three and eight fully armed paratroops. A civil version was produced as the AT-2 and was flown on the Chinese mainland by the Manchukuoan Aviation Company, the national airline of the Japanese puppet state of Manchukuo, and by China Airways Company serving occupied China. Some of these examples were picked up by the Chinese Communists at the close of World War II. In its civil configuration, the Ki.34, then identified as the Nakajima AT-2, carried eight passengers. Its top speed was barely over 200 mph. The Ki.34 received the wartime code name "Thora."

Nakajima Ki.43 Type 1 Hayabusa Fighter

Workhorse fighter of the Imperial Japanese Army Air Force in World War II, various models of the Nakajima Ki.43 Type 1 Fighter were spread out all over North China and Manchukuo when the war concluded, in September, 1945, and they ended up in Red Chinese service. The Japanese gave this aircraft the name Hayabusa, or Peregrine Falcon, a fighting bird in the hawk family. During the war it also received the U.N. Pacific code name of "Oscar." The Ki.43 was the first modern low-wing single-seat fighter with a retractable landing gear to serve in the Japanese Army Air Force. It entered production in 1941, thus receiving the Type 1 designation, because the Japanese based their service designations on the last digit of the current year. Over 3,200 of these nimble fighters were produced and it was still in production in its Model 3 version when the war ended. It was noted for its exceptional maneuverability, and was favored by many Japanese Army pilots over other Japanese fighters that later appeared in service. It was small and light with top speeds ranging from around 300 to over 360 mph in its latest models, although later Japanese Army fighters far exceeded it in performance. It was, in effect, akin to the American Curtiss P-40, German Me.109 and British Spitfire–all outstanding aircraft when they were introduced, but held in production far past their prime during the war years.

Nakajima Ki.44 Type 2 Shoki Fighter

Almost half engine, with a very small tail, the Nakajima Ki.44 reminds one of the flashy American and French racing aircraft of the late 1930's. The Japanese called it the Shoki, or Demon, and developed it for air defense duties as an interceptor. It therefore did not have long-range requirements; its function was to get off the ground fast, attack any raiding aircraft, and get back for another mission. The Allied Forces in the Pacific called it "Tojo," making it one of the few types to receive a name that wasn't American in origin. Shoki air combat regiments of the Japanese Army Air Force were stationed in central China and Korea, and quite a few found their way into Chinese Communist hands. It was a tough brute to fly. The type saw its primary use over Japanese home cities during the days of the B-29 bombings in 1944 and 1945. Span was 31 ft., 4½ ft. shorter than the Ki.43 Hayabusa, and yet the weight of the Ki.44 was somewhat greater than that of the Ki.43. Power was a Ha.109 radial engine of 1450 hp. Only an expert pilot could handle the Ki.44 Shoki because of its weight and high performance. During World War II, 1223 were built by Nakajima.

Nakajima Ki.84 Type 4 Hayate Fighter

Just about the toughest Japanese fighter to be met in numbers in World War II, the Ki.84 was called the Hayate, or

Hurricane, by the Japanese Army Air Force. The U.N. code name was "Frank." The Ki.84 was the ultimate development of the basic Ki.43 Hayabusa line and followed the same format, although it was unmistakably larger, bulkier, and stronger looking than its predecessor. The Hayate was difficult to fly, primarily because of mechanical problems. Hayate pilots knew they were flying one of the fastest and most powerful fighters in the Pacific theater of war, but they could never trust the performance of the aircraft. Engine failures and the tendency of the landing-gear legs to crack at unforeseen times kept the pilots on edge—and more often than not their worst fears were realized. Most of the Ki.84 production was in the last year of the war, and the material quality and workmanship were poor. But when a Hayate was in working trim it was hard to beat, and could outrun an American P-51D Mustang fighter at full throttle. Speeds of over 400 mph were clocked for the Hayate, a considerable achievement considering its heavy armament of four cannon. This gave it the fire power to knock down anything it could get close to. The mechanical problems in the air only compounded the maintenance problems on the ground, so it is easy to understand why the Chinese Communists did not make full use of their Hayate fighters in the early days of the civil war on the mainland. The Ki.84 had a 2,000 hp Nakajima Ha.45 radial engine, itself a touchy device, and a wing span just short of 37 ft.

North American P-51D Mustang Fighter

Regarded by many as the hottest American fighter of World War II, and well liked by its pilots, the North American P-51 Mustang was produced in a wide variety of models, of which the P-51D model, with its bubble canopy, had the largest production run. The P-51D, built at both the Inglewood and the Dallas plants of North American, had a production total of 7,956 aircraft. The power plant was a Packard

V-1650-7 liquid-cooled inline of 1,490 hp. Top speed was rated as 437 mph. The P-51D served in virtually every American theater of war. Its simple classic lines gained it admiration above and beyond its reputation as a killer. After the war the Mustang became one of the most widely exported American aircraft, serving in the air forces of dozens of nations around the world. Many Mustangs are still in service. The Nationalist Chinese had four fighter groups equipped with the P-51D by the end of 1947, making it the major fighter in the Nationalist Air Force during the civil war. An additional ninety-five were earmarked for the Nationalists in 1948, and ultimately reached them on Taiwan. By this time even the designation had changed, and with the adoption of the "F" prefix for "fighter" aircraft to replace the old "P" for "pursuit" from the days of the Army Air Corps, the aircraft became the F-51D. The last hasn't been heard from the F-51D, for now this veteran aircraft has been selected by the USAF for modification as a counter-insurgency fighter bomber. After about twenty years of retirement, old F-51D airframes are again being manufactured by Cavalier Aircraft Corp. in Florida and supplied to the USAF's Special Warfare Center at Eglin Air Force Base for testing and evaluation prior to delivery as "add-on" aircraft to foreign air forces already flying the earlier F-51D. The scheduled 1968 production rate for these new Mustangs is two aircraft per month. Although the wings have been strengthened and the vertical tail extended, the aircraft is unmistakably an F-51, giving the Mustang one of the longest fighter first-line service careers in history. It will be long remembered after its service days are over, even in Communist China, where a remarkably well preserved P-51D of the PLAAF is on display at the revolutionary museum in Peking.

North American B-25H and B-25J Mitchell Medium Bombers

The mid-wing B-25 Mitchell monoplane was the most widely used American medium bomber of World War II; almost 10,000 of these aircraft were built. The Mitchell first came into prominence in April, 1942, when it was the bomber used in the Doolittle Raid against Japan, taking off from the carrier *Hornet* in the Pacific. Mitchells were used in some numbers in China by the USAAF Fourteenth Air Force between 1943 and the end of the war. When USAAF stocks were turned over to the Nationalists at the end of the war, the Chinese Air Force picked up a substantial number of B-25H and B-25J models to add to the 43 B-25H Mitchell bombers already in Chinese service and many were later captured by the Chinese Communists during the Civil War. The B-25H was powered by two Wright R-2600-13 radials of 1,700 hp and had a top speed of 275 mph. It carried a crew of five. The B-25H model had a "solid" nose, replacing the glazed nose of the earlier B-25's, and mounted a heavy nose armament for use against ground and sea targets. The Mitchell was one of the few American aircraft to receive an enemy code name. Quite a few Mitchell bombers were shipped to the Soviet Union under Lend Lease in World War II, and many remained in Russian service into the 1950's. When the NATO code names were adopted, the Mitchells in Russian and Chinese Communist service were coded "Bank."

North American F-86F Sabre Fighter

Hero of the Korean War, if any mechanism could be, and leading jet fighter of the Free World for over fifteen years, the single-seat North American F-86 Sabre is one of the most famous fighter aircraft ever produced. License-produced versions were built in Canada, Australia, Japan, and Italy, and models of the Sabre served in over twenty air forces around the globe. Hundreds of the aircraft are still in use. The original F-86 design was based on a navy carrier fighter designed at the end of World War II. When captured wartime German data on the efficiency of swept wings in fighter aircraft became available soon after the war, North American modified the design to adopt the new wing form, and the F-86 was born. Its contemporary and prime adversary, the Russian Mikoyan MiG-15, utilized the same data. This situation added substance to the postwar joke between Russia and the United States that "our German engineers are better than your German engineers." As a result, the two aircraft were somewhat similar in appearance. The prototype Sabre first flew on October 1, 1947, two short years before the founding of the People's Republic of China, and barely three months before the prototype MiG-15 made its first flight in the Soviet Union. An advanced day-fighter model of the Sabre known as the F-86F first flew in the spring of 1952, ultimately becoming the most widely produced version of the series. It was powered by a J47 turbojet of 5,970 lb. thrust, giving the F-86F a top speed of 688 mph at sea level. Span of the F-86F is 37 ft. 1 in. Loaded weight is 15,198 lb. with an overload total of 20,611 lb. The F-86F quickly entered USAF service and by 1954 was authorized for export to friendly nations under the Mutual Defense Assistance Program. The F-86F soon became the leading American export aircraft, showing up wherever American influence and arms were welcomed. The Nationalist Chinese on Taiwan were among the first to receive the F-86F fighters under the MDAP; 320 of the day-fighter versions and seven RF-86F photoreconnaissance models reached the Chinese Air Force between November, 1954, and the summer of 1958. These were the aircraft that met the PLAAF over the Taiwan Strait in the air battles that took place in the summer and fall of 1958. By the time Captain Hsu Ting-tse defected to Peking with his Nationalist F-86F in the summer of 1963, the aircraft was of little importance to the Red Chinese. The Chinese Communists had long since captured, evaluated, and tested Sabre examples lost over North Korea in the early 1950's, and had learned about the capabilities of the F-86F model over the Taiwan Strait.

Northwestern Technological University Yenan No. 1 Utility Transport

A highly modified version of the Russian YAK-12 design, the Yenan No. 1 exhibits a number of creative design ideas. The high wing, cabin, and fixed-gear format of the YAK-12 have been maintained, but the Yenan No. 1 is a somewhat larger aircraft with a cabin for a pilot and up to five passengers. Its large smooth cowling and top speed of 121 mph suggest the use of a more powerful engine, possibly the Communist Chinese version of the 9-cylinder Ivchenko AI-14R of 260 hp. The Yenan No. 1 was built late in 1958 by the instructors and students of the Northwestern Technological University located at Sian, the provincial capital of Shensi province, one of the eleven multi-field science, engineering and technical polytechnic universities in operation in Communist China in 1958. Northwestern Technological University offered a series of courses in aeronautical engineering and followed the practice of an industrial on-the-job training "university" attached to a manufacturing facility. It was at this school that one of the earliest voices of dissent against the then current Chinese Communist practice of slavish dedication to Russian engineering and technology was heard. In 1957 engineering professor Liu Pu-tung complained that many Communist Chinese instructors taught from Russian texts and followed Soviet design practices without alteration, and thereby stifled student questions and originality. The suggested alternative of self-reliance and original thinking may well have led to the comprehensive changes so apparent in the design of the Yenan

No. 1. The design has been considerably cleaned up over the earlier YAK-12 and its other Communist Chinese derivatives. A single wing brace and single strut landing gear give it a modern look and suggest good STOL capability. The Yenan No. 1 was developed as a multi-purpose utility and light transport aircraft for agricultural work, flight-crew training, and sports flying. It would also have wide military application as a liaison and light cargo transport. Its name refers to the historic Red Chinese city of Yenan in upper Shensi province north of Sian, the mountainous capital of the Chinese Communists from the middle 1930's to 1947, when it was evacuated during the civil war.

Peking Aeronautical Engineering College Bei-Jing No. 1 Transport

Owing much to the design of the Russian Yakovlev YAK-16 transport, the Peking Bei-Jing No. 1 is a scaled-down version with less powerful engines and a nose wheel landing gear. Reportedly designed and built in 100 days by the teachers and students of the Peking Aeronautical Engineering College, the Bei-Jing No. 1 first flew on September 24, 1958. The Engineering College, with a five-year program in aeronautical engineering, has become a center for original Red Chinese thinking in aircraft design. A monthly publication called *Knowledge of Aeronautics* is published by the college and serves as the Communist Chinese equivalent to Western trade publications for aeronautics, although the Chinese version is notably lacking in advertising. This is one of the leading publications read in Chinese Communist aircraft-manufacturing circles, along with *Chinese Science*, the journal of the Chinese Academy of Sciences, which has an aeronautical engineering section under the dual editorship of Kuo Yung-huai, vice-chairman of the China Aeronautical Engineers Society, and Chien Hsueh-shen, director of the Institute of Mechanics.

The activities of the college reach into all areas of aviation and, as in other schools attached to manufacturing facilities, the Peking Aeronautical Engineering College involves its students in the actual work of aircraft construction both to train the future designers in the practice of aircraft production and to glorify the worker, who in Communist Chinese eyes is an equal partner in the creation of the country's aircraft. The Peking Bei-Jing No. 1 has a span of 57 ft. 5 in. compared to the 65 ft. 7 in. of the original Russian YAK-16. The smaller Chinese version is powered by two Chinese-built Ivchenko AI-14R radial engines of 260 hp each. In spite of the lower power, the Bei-Jing No. 1 carries a crew of two plus eight passengers at a top speed of about 185 mph, a capacity and performance just barely below that of the Russian transport. Originally described as a feeder transport for CAAC, the Bei-Jing No. 1 is available as a light cargo and troop transport as well as a multi-engine pilot and crew trainer. It is a relatively unsophisticated and attractive design and one that would find useful work in any modern air force. With the coming of the Cultural Revolution the Peking Aeronautical Engineering College was closed, along with the other specialized and advance training schools throughout the mainland, and many of its students became Red Guards. Early in July, 1967, the College was reopened to become one of the first, perhaps actually the very first, among the schools in China to get back into operation after being closed for over a year. This may indicate the school's importance to Red China's future in the air.

Peking Aeronautical Engineering College
Sha-Tu No. 1 Light Transport

On March 15, 1958, the prototype of a small Russian twin-engine high wing STOL transport designed by Oleg Konstantinovich Antonov made its first flight in the Soviet Union.

It was designated the AN-14 and was affectionately called the Pchelka, or Little Bee. It was designed as yet another replacement for the transport version of the Polikarpov PO-2 and could carry eight passengers. Less than a year later, a very short time by aeronautical standards, the Chinese Communists at the Peking Aeronautical Engineering College, wheeled out what appeared to be a scale model of the same aircraft, a new version that carried only four passengers. It was obvious that the Russians had shared one of their latest design ideas with the Chinese Communists as soon as they had the aircraft in usable form, and once again the Chinese engineers and designers had scaled down a Russian type to fit their needs, demonstrating a design sophistication far beyond that anticipated by the outside world. The Chinese Communists claimed that the students and teachers of the Peking Aeronautical Engineering College had created the original design and then built it in the record time of sixty-eight days. They also announced that the type would be mass-produced at the Peking production facility as the Sha-Tu No. 1, or Capital No. 1, a rather strange name for a Chinese Communist aircraft. If the Sha-Tu No. 1 had been an exact duplicate of the Russian AN-14, the sixty-eight-day record would have been an amazing feat in itself, but the fact that the Communist Chinese model was a complete redesign with its prototype already completed in February, 1959, makes the Chinese accomplishment, if true, a remarkable one. The Sha-Tu No. 1 is powered by two Chinese-built Mikulin M-11FR radial engines of 160 hp each in the same helmeted cowlings used on the Chinko No. 1 and the Hiryu No. 1. Assuming that the Chinese Communists had the benefit of the original Russian blueprints for the AN-14, they still had to re-engineer the entire airframe completely, for the wing structure of the Chinese model was considerably smaller than the Russian version, which was powered by two 260 hp Ivchenko AI-14R radial engines. The instructors and students had the further handicap of working under semi-primitive conditions. As one of the newly established specialized engineering schools set up by the Chinese Communists in the 1950's, the Peking Aeronautical Engineering College was just getting into operation at the time the Sha-Tu No. 1 was being developed, and the school's facilities at the time could be classified as good but far from adequate. Communist Chinese announcements state that the Sha-Tu No. 1 was specifically created for the Special Flight Group for commercial and agricultural use, although an aircraft of this type would also have wide military application with its rough field STOL capability. The larger Russian version received the NATO code name "Clod." The Chinese version is apparently yet to be named.

Peking Aeronautical Engineering College Red Banner No. 1 Utility Transport

The honor of being the first completely original Chinese Communist aircraft design to reach the prototype stage may belong to the Peking Red Banner No. 1, a large STOL utility transport monoplane in the class of the De Havilland-Canada DHC-2 and of the Russian Antonov AN-2 biplane. It is possible that the Red Banner No. 1 design is actually based on the AN-2, for some of the Russian transport's lines can be discerned in the nose and cabin outlines, but otherwise the Red Banner No. 1 is very much its own aircraft. A high-wing monoplane with a bulky forward fuselage for passengers and cargo, followed by a long boom supporting the tail assembly, the Red Banner No. 1 is a large aircraft that sits high on the ground. The use of a single brace-supported wing makes it look somewhat more modern than the AN-2 design although it appears to be powered by the same type of engine, a Communist Chinese version of the Russian Shvetsov ASh-62-IR 9-cylinder radial of 1,000 hp, possibly the Chinese Model 670-13 as used on the Shenyang Fong Shou No. 2, the Chinese version of the AN-2. The multi-purpose Red Banner No. 1 was developed for a wide range of duties, from civil and military cargo and passenger transport to agricultural roles as a crop sprayer and duster. The oversized fixed landing gear makes it adaptable to rough field operations, and the number of cabin windows indicates it can carry from six to eight passengers. The Red Banner No. 1 was produced by the Peking Aeronautical Engineering College late in 1958 in parallel with the Shenyang production of the Fong Shou No. 2, although little has been heard of the Red Banner No. 1 design since then. The profusion of Red Chinese aircraft under development for production during this period may have led to a reduction in model types and the dropping of the Red Banner No. 1 design, although aircraft in this class would have been invaluable to the Chinese Communists during the trying years of the early 1960's. Unless the Red Banner No. 1 was plagued with an insurmountable number of mechanical bugs, it would not be surprising to see this type become one of the most widely produced utility types on the mainland.

Petlyakov PE-2 Attack

One of the surprises facing the Germans after their first military successes in Russia in 1941 was the vitality of the various elements of the Soviet Air Force, as well as Russian aircraft of modern concept, particularly in the ground-attack classification. German columns found themselves being assaulted by heavily armed aircraft specifically created to meet the threat of surface armor, including the brutish-looking Ilyushin Shturmovik series of single-engine assault planes and the attractive twin-engined low-wing Petlyakov PE-2 monoplane dive bomber with a crew of three, which had first flown in prototype form in 1939 and had entered production in June 1940, only a year earlier. Combat experience led to the PE-2FT model, which had two forward firing machine guns, a 12.7 mm gun in a dorsal turret, another under the fuselage facing to the rear, and two side-mounted 7.62 mm guns. The PE-2 survived the war essentially unchanged, to become one of the Russian types available in quantity for export in the immediate postwar period. Its dive-bombing days were over, but it took on new life as an advanced trainer to train crews for the larger and newer Tupolev TU-2 attack bomber, another wartime type widely exported to Soviet satellite nations after World War II. The PE-2 was designed by Vladimir M. Petlyakov, who was a close associate of Andrei Nikolaievich Tupolev, the dean of Russian aircraft designers and the Soviet Union's leading bomber designer. Petlyakov would probably have been very influential in the design of Russia's early jet bombers following the war if he had lived, but an air crash took his life in 1944. His PE-2 design remained as his monument, showing up in the air forces of Yugoslavia, Czechoslovakia, Poland, and Communist China in the late 1940's and early 1950's. Top speed of the early-production PE-2 was 335 mph with two 1,100 hp Klimov VK-105R liquid-cooled inline engines. Its diving speed, held in check by dive brakes on the wings, was 373 mph. Empty weight was 12,900 lb. with a loaded weight of 16,930 lb. and up to 18,730 lb. when overloaded. Early in 1943 a modified model of the PE-2FT began to enter service powered by two VK-105PF engines of 1,210 hp, pushing the top speed slightly over 360 mph. It was this model that was sent to Red China. NATO code name for the PE-2 is "Buck."

Polikarpov R-5 General Purpose

The dramatic expansion of the Soviet aircraft industry based on indigenous Russian designs in the late 1920's and early 1930's had a profound effect on Russian foreign policy. The Russians carefully studied contemporary foreign designs and created aircraft similar in concept, finally giving the Soviet

Union complete command over an arms-producing facility that could supply the defensive needs of the leading Communist state while also providing aircraft and arms for other Communist movements in Europe and Asia. The Polikarpov R-5 was one of the first of these new types to reach production, and was developed in 1927 as a two-seat biplane replacement for the older Aviakhim R-1.M-5, then reaching the limit of its service career. The reliable R-5 became the standard reconnaissance bomber and general-purpose type of the Soviet Air Force and remained in production for almost ten years in a wide variety of models. Russian reliance on German technology was evident in the R-5 design. The aircraft was similar to the German Heinkel series of general-purpose biplanes of the same period, and the 680 hp M-17 power plant was a license-produced version of the German BMW VI liquid-cooled inline engine. This was the same engine utilized by Heinkel in some models of the corresponding German aircraft. The R-5 followed Russian involvement in Europe and Asia, showing up on the Loyalist side of the Spanish Civil War and, clandestinely, with the Red Army of China in North Shensi in 1937. The secret was well kept. Although the Nationalists claimed that the Chinese Communists were receiving Russian aircraft, no contemporary reports confirmed the fact. It has only been in late years that the Chinese Communists admitted receiving such aid. The Polikarpov R-5 was a large aircraft, with a span slightly over 50 ft. 10½ in., a length of 31 ft. 10 in., and a loaded weight of 6,160 lb. Many R-5 aircraft and their variants were in Russian service at the time of the German invasion in 1941, seeing action against the German Army as it surged through eastern Russia. The Germans saw the use of the obsolescent R-5 biplane as a demonstration of Russian backwardness.

Polikarpov PO-2 Utility

Without question the oldest aircraft type still active in general use to this day, the ancient-looking Polikarpov PO-2 open cockpit biplane was first introduced to the world at a European air show in 1927. It was one of the very first original Soviet aircraft to be exhibited outside the Soviet Union. Its designer was Nikolai N. Polikarpov, later to become famous for his original Russian fighter designs in the early 1930's. The PO-2 was destined to become one of the most prolifically produced aircraft ever built; well over 20,000 of them were constructed between 1927 and 1948, a production run of over twenty years. The PO-2 started out its service life as the U-2 Primary Trainer for the then infant Soviet Air Force, replacing the earlier Aviakhim U-1 Avrushka trainer. By the middle 1930's the U-2 was produced as a reconnaissance and liaison plane, an air ambulance with two stretcher-carrying pods on its wings, and as an ambulance and light transport with a long cabin built lengthwise over its fuselage behind the pilot's open cockpit. It was the best known utility aircraft in Russia in the years before World War II and was allocated in great numbers to agricultural spray units and civil sports clubs, and to glider schools for use as a glider tug. The U-2 earned a new lease on life with the German invasion of Russia on June 22, 1941, and soon showed up over the fighting front as a light military transport, as a low-level attack bomber and, of all things, a night bomber. In 1944 the U-2 was redesignated as the PO-2 to give credit to its designer in line with other Russian designation changes. The end of the war found the PO-2 being liberally distributed to the new Soviet satellite nations. Even the Korean War front felt its sting, if it could be called that, when the venerable PO-2 was used in "Bedcheck Charlie" raids at night over American positions, making an ungodly racket and dropping bombs now and then to keep the American forces from sleeping. Old as it is, the PO-2 can be expected to remain in use throughout the Communist world for years. Examples are flying in Europe and in mainland China with a wide variety of new engines, although the production models most often in use mount a 160 hp 5-cylinder Mikulin M-11FR radial engine that give the standard PO-2 a hard-to-believe top speed of 96 mph. Red Chinese examples are in widespread use for crop spraying, and serve as glider tugs for the many civil and PLAAF glider schools. In honor of its long life, and the fact that the aircraft is still in common use, the PO-2 received the NATO code name "Mule." It seems fitting.

Republic P-47D Thunderbolt Fighter

Affectionately known to its pilots as "Jug," the Thunderbolt was big, heavy, and brutish. Late versions of the "D" model introduced the bubble canopy to the P-47 series, making the P-47D model in all its variations the most widely produced version (over 12,000 aircraft). Power was, for its time, a massive 2,000 hp Pratt and Whitney R-2800-21 air-cooled radial, giving the P-47D a top speed of 420 mph, just slightly slower than its co-fighter, the North American P-51D Mustang. Deliveries of the P-47D to the USAAF in China began in April, 1944; they were to be used to defend the B-29 bomber bases in China against Japanese aircraft. The China-based Fourteenth Air Force didn't like the P-47D because its engine burned up 50 per cent more gas than the lighter Mustang, and all aviation gas in China during the war years had to be flown over the "hump" from India. The assignment of the "Jug" to China astounded the American pilots there, and yet the problem was only compounded after the Pacific war ended. As soon as the fighting was over, the Chinese Nationalists soon purchased the Thunderbolts from the USAAF, picking up three squadrons of P-47D-21RE, P-47D-23RA, and P-47D-35RA and a few of the faster P-47N fighters for use on the mainland, and eventually in the civil war. Additional Thunderbolts were later shipped in from the United States, and many of them went with the Nationalists to Taiwan, while the remainder were left on the mainland to be captured and enter service with the air arm of the PLA.

Shenyang YAK-18 Primary Trainer

Only by producing its own aircraft power plants can a nation become truly self-sufficient in the air. The Chinese Communists were quick to recognize this fact. They obtained licenses to produce Russian engines along with aircraft and established the Institute of Mechanics early in 1956 in order to engineer and produce their own power plants. Communist China started modestly by first producing the 5-cylinder Mikulin M-11FR of 160 hp, a relatively simple engine by modern standards. By 1958 they were ready to attempt the production of the far more complex power plants needed for modern aircraft. In rapid succession the 9-cylinder Ivchenko AI-14R of 260 hp, the 9-cylinder Shvetsov ASh-62-IR of 1,000 hp, and the Klimov RD-45 and VK-1A centrifugal-flow turbojets—the last being for China's Shenyang fighters and jet trainers—entered production. Communist China's ultimate break with the Soviet Union revealed the wisdom of these moves, for theoretically China could now sustain her own aircraft production without total dependence on the outside world. Production of the M-11FR enabled the Chinese Communists to develop a broad range of original prototypes. Its first use was for a Chinese version of the YAK-18 primary trainer. The original production run of these aircraft made use of Russian airframes, components, and parts assembled in China at the production facility near Shenyang that was to become the National Aircraft Factory, with power soon supplied by the Chinese built M-11FR. By mid-1954 the Chinese Communists had apparently cut the bond and the Shenyang YAK-18 was produced in its entirety by the Shenyang facility. Peking radio reported that the first aircraft to be completely produced in Communist China made its initial test flight on July 26, 1954. The Chinese suggested that the aircraft was a primary trainer of original design, but production models revealed it to be the license-produced YAK-18. In its original version the empty weight was 1,795 lb., and the aircraft had a top speed of 153 mph. Availability of the more powerful Ivchenko AI-14R engine in Communist China suggests that the Chinese may also modify and further develop the Shenyang YAK-18 in the same manner as the Soviets have done with their basic Yakovlev YAK-18 design over the years. Code name is "Max."

Shenyang MiG-17 Fighter

Pride of the Communist Chinese aircraft industry in the late 1950's and early 1960's, and the first truly modern military aircraft produced in the People's Republic of China, the Shenyang version of the Russian MiG-17 was a license-produced version of the Mikoyan MiG-17f Day Fighter. It is the equivalent of the NATO-coded "Fresco-C" version of this versatile fighter. The Shenyang MiG-17 firmly established Communist China as a producer of advanced aircraft. A complete manufacturing facility was established at Shenyang, Manchuria, on the site of the former Manshu Airplane Manufacturing Company, previously dismantled by the Russians following World War II. Under Russian direction groundwork began late in 1954 on the National Aircraft Factory of the People's Republic of China for the assembly of Russian-made MiG-17 components. The power plant was a Russian-made Klimov VK-1A centrifugal-flow turbojet of 5,950 lb. thrust with a boost to 6,990 pounds with afterburning. The first Chinese assembled MiG-17's were completed in 1956, and by the middle of 1957 production had moved up to fifteen finished aircraft a month using Russian engines, instruments, radio gear, and other components. Throughout this period the factory was being converted to Chinese airframe production with continued use of Russian equipment, and by the end of 1959 the National Aircraft Factory was completing twenty-five aircraft per month. The final step was total Communist Chinese production of engines, instruments, electronic gear, and armament. The Shenyang MiG-17 was still being produced in 1962, although it was then being phased out of production by the newer Shenyang version of the MiG-19. Standard Shenyang MiG-17 armament was three nose cannon.

Shenyang MiG-15UTI Fighter Trainer

When the prototype of the license-produced version of the Russian Mikoyan MiG-15UTI two-seat jet conversion trainer was rolled out of the National Aircraft Factory at Shenyang and flown on September 30, 1958, the Communist Chinese press introduced it as the first jet aircraft of original Chinese design. It is likely that the Chinese Communists made a few changes in the basic Russian design which, to their way of thinking, gave them the right to claim the aircraft as their own. Such statements must have annoyed the Russians, who, by that time, were already having second thoughts about their commitment in China, having just denied the Chinese Communists new aircraft and weapons which the latter intended to use in their planned invasion of Taiwan. The Shenyang MiG-15UTI reportedly had a ceiling of over 45,900 ft. and a top speed of 620 mph with a Communist Chinese–built Klimov RD-45 turbojet of 5,450 lb. thrust. While this claimed performance at least matched, and probably exceeded, the Russian MiG-15UTI trainers produced in the Soviet Union, it was still within the realm of possibility. The Shenyang versions were somewhat lighter in weight than the Russian versions and did not carry the two 23 mm cannon so frequently found on the examples produced in the Soviet Union or under license in Czechoslovakia or Poland. The production line for the Shenyang MiG-15UTI was set up parallel with the Shenyang MiG-17 line at the National Aircraft Factory. By the middle of 1959, production of the Shenyang MiG-15UTI was well under way and the aircraft was becoming the standard jet conversion trainer of the PLAAF, eliminating Red China's dependence on the Soviet Union for trainers of this type. The Shenyang MiG-15UTI remained in production in Manchuria until well into the 1960's.

Shenyang AN-2 Fong Shou No. 2

It was a proud day for the aircraft industry of the People's Republic of China when, in December, 1957, the first Shenyang-produced AN-2 utility biplane was rolled out of the factory for its press introduction. It was announced as Red China's first locally produced civil aircraft with intended use for CAAC transport, crop spraying, aerial survey, and pilot training. It became the aircraft type most frequently associated with the aircraft industry of mainland China, and it was a good choice for Communist Chinese needs. Produced under a Russian license, and virtually a direct reproduction of the Antonov AN-2, many did go into civil service in a wide variety of duties. But the Shenyang AN-2 was also delivered to the PLAAF in some numbers for use as an operations trainer, cargo transport, and utility type. The Communist Chinese prototype first flew early in 1958, powered by an imported Russian Shvetsov ASh-62-IR radial engine of 1,000 hp. The initial Chinese production models made use of imported engines, but later production was powered by the Chinese Model 670-13, a 9-cylinder radial of 1,000 hp developed by the Institute of Mechanics based on the Russian ASh-62-IR power plant. In Chinese service the Shenyang AN-2 was called the Fong Shou No. 2. It carried a crew of two and eight or ten passengers, or a cargo load of 2,700 lb. When production got under way early in 1958 the Chinese Communists found they were able to produce the type with less difficulty than expected, and by the end of the year production forecasts were increased by 50 per cent over the original production schedules. Span of the Shenyang Fong Shou No. 2 is 59 ft. 7½ in. Cruise speed with the Chinese-built Model 670-13 engine is 110 mph. The NATO code name "Colt" also applies to the Communist Chinese version of the AN-2.

Shenyang Whirlwind-25 Helicopter

When the Shenyang Whirlwind-25 helicopter was introduced to the Red Chinese press late in 1959, it was said to be the first original Chinese helicopter design. Typically, this description was wrong on two counts. Back in the days of the Nationalist Government on the mainland, starting as early as 1944, Major General Chia-jen Chu, an officer in the Nationalist Air Force, designed and built a series of light helicopters for possible use by the Nationalist Army and Air Force. With the loss of the mainland, General Chu took his designs with him to Taiwan and continued his work there. On the second count, the Red Chinese were equally incorrect. A close study of the first photographs of the Whirlwind-25 revealed that this Shenyang product was a very close copy of the Russian Mil MI-4 helicopter. It appeared that the only change was a new Communist Chinese name and designation. The Communist Chinese MI-4 was first described as a civil type for air-taxi, agricultural, and special-flight-group work. The demonstration model was painted in civil colors and decorated with flashy trim lines, and the covering press stories suggested that the Whirlwind-25 would soon be available in great numbers. It was, but the bulk of the production went to the PLAAF. The Russian version is powered by a Shvetsov ASh-82V radial of 1,700 hp, and it would appear that this power plant is produced in Communist China. Agricultural models of the

Whirlwind-25 carry a bin for crop dusting and chemical sprays. Loaded weight is in the neighborhood of 15,875 lb., the weight of the Russian MI-4. The Shenyang version is so similar to the Russian model that the code name "Hound" still applies.

Shenyang Chinko No. 1 General Purpose

Production of the 5-cylinder Mikulin M-11FR radial engine in Communist China under a Russian license gave the Chinese a reliable 160 hp power plant that could be used on a wide variety of light aircraft types. One of the first attempts was a Communist Chinese version of the basic Russian Yakovlev YAK-12 design, a high-wing monoplane that had the look of an American light plane. Called the Chinko No. 1, the prototype was built in 1958 in two and a half months at the Shenyang Aviation School, an aeronautical trade school attached to the National Aircraft Factory, by the school's teachers and students aided by local labor. The Shenyang version of the YAK-12 was outwardly similar to the Russian model, an aircraft that was first introduced over ten years earlier. The Chinko No. 1 was actually a transition model, for while it used the M-11FR power plant of the older Russian model, it incorporated some of the design and construction improvements of the later Russian versions. In appearance it was a braced high-wing monoplane with a cabin for its pilot and a capacity of three passengers. Some confusion has existed over the name of the school producing the aircraft, with alternate translations describing it as the Shenyang Aeronautical Engineering College as well as the Shenyang School of Aeronautical Industry, although its trade school aspects seem to favor the latter version. This school was closed during the Cultural Revolution along with the other trade and higher educational schools throughout Red China. The Chinko No. 1 has a wing span of 41 ft. 4 in. and a length of 28 ft. 11 in. Wing area is

256.7 sq. ft., and the top speed 121 mph. The NATO code name "Creek-A" applies to this aircraft.

Shenyang MiG-19 Fighter

The last words heard from two U.S. Navy A6 attack planes from the carrier *Constellation* after they were forced over the Communist Chinese border on August 21, 1967, were "Farmers, Farmers." Within hours the Chinese Communists reported they had been shot down over Kwangsi province. At the eighteenth anniversary celebration of the founding of the People's Republic of China in Peking in October, 1967, the skies were filled with PLAAF fighters, with silver Shenyang MiG-19 fighters flashing overhead. The Communist Chinese version of the MiG-19 "Farmer" is very much in evidence, and promises to be an adversary for years to come. When construction of the Russian Mikoyan MiG-19 design was authorized for production at Shenyang under license in February, 1959, the Chinese Communists were exhaulted. It was a clear demonstration, or so both sides believed, of the eternal permanence of the Sino-Soviet partnership. But construction of Communist China's first supersonic fighter aircraft proved to be a nightmare. The ability of the Chinese Communists to survive the cutting off of Russian aid following the Sino-Soviet split, and the ultimate production of the Shenyang MiG-19 at the National Aircraft Factory under completely Chinese direction, was the making of the aircraft industry on the mainland. The Shenyang MiG-19 was based on the early MiG-19 model, with power supplied by two Chinese versions of the Klimov VK-7 turbojet reportedly delivering 6,170 lb. of thrust each. These new turbojets were produced at the second Ministry of Machine Building facility that produced the earlier Klimov VK-1A jets for the Shenyang MiG-17. With the coming of the Shenyang MiG-19, this productive capacity was virtually cut in half, for now the Chinese Communists were producing a fighter that needed two power plants for each airframe. The American attitude toward aircraft produced in Communist China seems to demand that they be poorly assembled and unreliable. This appeared to be confirmed by a report credited to the Pakistani Air Force, one of the few foreign air arms to receive the Shenyang MiG-19, which suggested that the Chinese product was "technologically crude." Pakistani technicians were quoted as remarking that the parts of the Shenyang MiG-19 had been poorly machined and that overall assembly was irregular, with the finish often imperfect and bumpy. But these comments are completely belied by reported discussions with Pakistani pilots, who openly state that the Chinese fighter is superior in a number of ways to the American aircraft in their air force. They report that the finish of the Shenyang MiG-19 surpasses that of the American-built F-86, and that the electronic-component reliability matches that of their American equipment. They also credit the Shenyang MiG-19 with outstanding performance and regard it as an easy aircraft to handle in the air. Although Pakistan does not seem to have reordered the type after the initial Chinese delivery of eighty aircraft and is now shopping for more modern European aircraft, Pakistani experience with the type could well pave the way for Communist China's further export of its advanced aircraft. Loaded weight of the Shenyang MiG-19 is reported to be slightly over 18,000 lb., with dimensions the same as the Russian model.

Shenyang MiG-21 Day Fighter

When fighting finally broke out during the border war between India and Communist China late in 1962, the Western Powers, as well as the Indians, chortled over the fact that the Soviet Union did not agree to Red China's demands for MiG-21 fighters while they proceeded with plans for production of the aircraft in India by Hindustan Aircraft, Ltd. Most of the world saw this as a sign of possible disagreement between the two major Communist powers, and the Russian intransigence was considered one of the first overt indications of a break in relations between Russia and Communist China. What was not known at the time was that the Russians had previously committed the MiG-21 to China, and then backed out of the agreement after the Sino-Soviet split had become a reality. Russian refusal to send more MiG-21's to China forced the Chinese Communists to face a dangerous and embarrassing situation. At just about the same time the Chinese Communists made up their minds to embark on a program of self-reliance and produce a Shenyang version of the MiG-21 using as patterns the few examples of the Russian fighters then in their hands. The secret Chinese decision to proceed with the development and production of the most sophisticated aircraft yet to be produced in the People's Republic of China was an expensive one, for it required the re-creation and production of a 9,500 lb.-thrust axial-flow turbojet, a highly strengthened airframe for supersonic flight, original instruments and electronic gear, and the original development of all the supportive training and maintenance equipment needed by an aircraft designed to fly in the Mach 2 range. The Chinese Communists claim that they were able to complete the task in five years, suggesting a Russian delivery date of late 1959 for the original pattern aircraft. The Communist Chinese have been scrupulously careful not to let one of their MiG-21 fighters fall into Western hands, and only on rare occasions have any of these aircraft been flown out of the confines of mainland China. Based on the original MiG-21 examples supplied to Communist China, the Shenyang MiG-21 is assumed to be equivalent to the NATO-coded "Fishbed-B" version of the Soviet fighter. This would make it less powerful and somewhat slower than its newer Russian-produced counterparts, although the Chinese are thought to have improved on the original Russian version. The Shenyang MiG-21 and its follow-on variants are expected to be important PLAAF fighters for many years to come. Communist China's desperate need for new fighters and the difficulty of altering production of an apparently successful aircraft under the bureaucratic government of the People's Republic of China may force continued production of the type far beyond its useful years. The temptation to modify, improve, and innovate new models utilizing the same basic airframe in order to maintain effective production will be hard to avoid. This would start the vicious circle of much-too-much and much-too-late experienced with the Shenyang MiG-17 all over again for the PLAAF.

Shenyang Multi-purpose Fighter

Cut off from both the Soviet and the Western world in the early 1960's and faced with enormous production problems with aircraft already on the lines at Shenyang, Communist China could not be expected to develop and produce modern fighter aircraft with parity to that of the world's leading air powers. But now many of these conditions have changed. Powerful domestically designed Chinese turbojets are now in production in Communist China. The Chinese have had an opportunity to study, fly, test, and dissect some of the most modern jet fighter aircraft in the world, including Russian-built MiG-21f and MiG-21Pf fighters, as well as American McDonnell F-4 Phantom II and Republic F-105 fighters and fighter bombers downed over North Viet-Nam and the Chinese mainland. Communist China has already demonstrated that her engineers are capable of duplicating and innovating new fighter designs. With the considerable amount of new design work now taking place in Communist China, the introduction of her next fighter model will probably be a rude shock to the whole world, and an achievement of considerable pride to the Chinese. It is the stated policy of the Military

Science Academy to develop reliable aircraft types suitable for immediate production by the seventh Ministry of Machine Building at the National Aircraft Factory rather than devote engineering time to new design projects that may require years of development and testing time. As a result, original design work has a very low priority while the improvement and innovation of existing types are actively pursued. The format for a major new fighter design already exists in Chinese hands. Modification of the existing Shenyang MiG-21 airframe to mount two of the new Communist Chinese 9,500 lb.-thrust turbojets in the fashion of the McDonnell Phantom II would give the Chinese a multi-purpose fighter comparable to the leading fighters in both the Soviet and the American air forces. Performance in the Mach 2 range can be expected. This potential new Red Chinese aircraft may already be flying.

Shenyang Medium Bomber

Preliminary plans for Communist China's Shenyang Medium Bomber design were in work as early as January, 1965, when the production of the type by the seventh Ministry of Machine Building was authorized at a meeting of the National Defense Council in Peking. The apparently long delay in introducing this aircraft is not surprising, for it is possible that the creation and construction of a sophisticated jet bomber of this type would be just as difficult and expensive for the Chinese Communists as the development of their thermonuclear bomb, and in some areas perhaps even more so. The Chinese Communists obviously need help, for they have been bidding high for German airframe designers and engineers, directly competing with American firms attempting to recruit the same group. Many of these designers and engineers worked on aircraft and missile projects in Egypt, and were recalled to West Germany after Nasser's recognition of East Germany in 1965. Cancellation of aerospace projects in West Germany left many of them jobless. Long-term contracts, payments in Western currency in Swiss banks, and liberal expense benefits were offered by the Chinese Communists as incentives to Germans experienced in multi-jet, STOL, and VTOL (vertical-take-off-and-landing) aircraft. As such engineers go to Red China, the initial influx of new ideas, with the corresponding problems in language, may well be a deterrent to smooth and rapid progress on the Shenyang bomber. But once this period of confusion has been overcome, the advantages of the experience can be felt. Expert opinion early in 1967 estimated that Communist China would not put an indigenous long-range bomber force into service until the late 1970's. These opinions are now being drastically revised, and it is believed that Communist China has given a top priority to the development of a strategic jet bomber force. Present thinking favors the availability of a 2,200-to-2,800-mile range bomber by 1970, and these bombers may be available in regimental strength a few years later. Communist China is already producing turbojets estimated to deliver 9,500 lb. thrust for the Shenyang MiG-21 fighter. Four of these existing jet engines would give the Chinese bomber a thrust equal to the Russian Tupolev TU-16 or the American Convair B-58 Hustler, certainly not the latest types in Russian or American service but both advanced bombers in their own right.

Sikorsky R-6 Helicopter

The first USAAF helicopter to reach a production run of over a hundred units, the World War II Sikorsky R-6 was designed by Sikorsky and built by Nash-Kelvinator; actual production took place from 1943 through 1945. Major production model was the R-6A, of which 193 were built. The Franklin O-405-9 engine turned out 240 hp, which was enough to give the R-6A a top speed of 105 mph. Following the fabric-covered R-4B helicopter into wartime service, the

R-6A was the first all-metal combat helicopter available in quantity, and it pioneered the use of the helicopter in rescue missions and communications. The delivery of Sikorsky R-6A helicopters to Nationalist China had not been reported in the press, and thus the appearance of a captured example in Peking at an aircraft display in the spring of 1965 came as a surprise to Western observers.

Sud-Aviation SE-3160 Alouette III Helicopter

Currently one of the most successful and widely used helicopters in the world, the French Sud-Aviation Alouette III has been produced in great numbers and has been widely exported to both non-Communist and Communist nations. It holds the world altitude record for helicopters. This high altitude capability attracted both Communist Chinese and Indian interest following the border war between these neighboring nations. Late in 1963, Red China investigated the possibility of obtaining a license to produce the Alouette III at Shenyang, but the agreement did not materialize and the idea was shelved in 1964. About this time India also investigated the possibility of producing the Alouette III for her own use. Arrangements were made to deliver twenty French examples and produce eighty more at a special plant to be set up in India under French supervision. Communist Chinese interest did not die, however, for on March 24, 1967, the French announced in Paris that the People's Republic of China had purchased fifteen examples of the Alouette III, all to be delivered before the end of the year. No mention was made of possible Chinese production, but past Red Chinese interest suggests the very real possibility of a Shenyang version, perhaps to replace the Shenyang Whirlwind-25 on the Chinese helicopter-production lines. The French are obviously eager to license the production of the Alouette III, and as soon as the arrangements for the production of the type in Israel in 1968 are completed, the French licensing team will be looking for new customers and assignments. Shenyang production would give the Chinese Communists a helicopter that could operate efficiently in the thin atmosphere of the Himalayan region bordering India, a capability beyond that of the Shenyang Whirlwind-25. Communist Chinese production would also create the need for a code name, the first for a French design. The Sud-Aviation version is designated SE-3160, and carries seven, including the crew. Power is a Turbomeca Artouste III-B turbine of 550 hp. The main rotor diameter is 36 ft. 1⅛ in., and the fuselage is 33 ft. 4⅛ in. long. Loaded weight is 4,630 lb. Top speed is 130 mph, and cruising speed is 118 mph.

SZD IS-3 ABC Primary Glider

The world doesn't return to normal overnight when a major war ends, particularly for those nations deeply involved in the conflict, which bear the scars of combat. But the end of a war does suggest that new conditions will evolve and perhaps the terror will end. In short, hope for a better life returns, whether or not the facts bear this out. The end of World War II in Europe left the continent battered, poor, frightened—and hopeful. Even the most fought-over nation of the war, and the country in which the war began, soon showed feeble signs of recovery. Thus in Bielsko-Biala, Poland, in April, 1946, a group of gliding enthusiasts formed the Institute Szybownictwa, or Gliding Institute, a scant year after the collapse of Nazi Germany. With the nationalization of Polish industry under its postwar Communist government, the Gliding Institute found itself responsible for the development and production of Polish training gliders and sailplanes. One of the Institute's first projects was the creation of a simple open-frame primary glider built along the lines of the original German primary gliders of the late 1920's, a classic format produced and used wherever gliders were flown in the world.

Towed into the air by a winch or an automobile, but in immediate postwar Poland more often than not by a crew of running helpers who pulled the glider into the air by hand, the Institute's powerless primary glider carried its single passenger a short distance above the ground, leaving further flight and safety in the pilot's hands. Successive development led to the successful third design project of the Institute, which was produced as the IS-3, or Gliding Institute No. 3, fifty of which were sold and delivered to Communist China in 1955 and 1956. The designers gave it the name "ABC" to note that it was a primer in the art of gliding.

SZD IS-3 ABC-A Primary Glider

The successful SZD IS-3 ABC led to a further development of the Polish type which entered production as the Model A, or IS-3 ABC-A. These gliders were built at Bielsko-Biala on the grounds of the Gliding Institute, and the new primary glider soon became one of the most sought-after glider trainers in both Communist and non-Communist Europe. Before World War II, Poland had been one of the leaders in the sport of gliding, and this reputation was quickly being rebuilt. Poland soon found herself in the aviation export business, and the People's Republic of China was one of the major customers. Fifty examples of the IS-3 ABC-A were sold to Communist China along with other Gliding Institute products. The IS-3 ABC-A was of a simple layout, carried only its pilot, and was able to be towed at a speed of 75 mph. Its span was 29 ft. 6¼ in., and the length was 20 ft. 6¾ in. The glider weighed only 230 lb., and had an average loaded weight of about 400 lb. with its pilot strapped into its slab seat.

SZD Salamandra 53A Primary Trainer

One of the first gliders to enter production at the Polish Gliding Institute after World War II was the Salamandra

53A primary glider. It had an open frame like the IS-3 ABC but also had a small cockpit pod for its pilot. The Salamander, the English translation of the very similar Polish name, was actually an old design well known in pre–World War II gliding circles. The prototype had been built in 1936, and because the design was tried and true it entered limited production at the Gliding Institute in 1946 as the improved Model A, or 53A. It was one of the most popular of the primary gliders produced by the Institute and remained in production for many years. In 1955 a production series was built for export to the People's Republic of China and over fifty were ultimately shipped there. Somewhat larger and heavier than the IS-3 ABC-A, the Salamandra 53A had a span of 40 ft. 11½ in. and was 21 ft. 3¼ in. long. It weighed 308 lb. empty and about 496 lb. with its pilot.

SZD IS-4 Jastraab Sailplane

Soaring has been described by its advocates as the most exciting sport conceived by man. Its challenge, the freedom of the silent air, the self-expression and personal achievement that come from a successful flight, and the luxury of loneliness and closeness with nature have made soaring popular the world over. It is particularly popular in the Soviet Union, the Polish People's Republic, and the People's Republic of China. In such controlled societies one can understand why. When Communist China first invited Polish technicians to the mainland to organize a gliding movement in China, one of the initial gliders that accompanied the mission was the single-seat IS-4 Jastraab sailplane. It wasn't very large as sailplanes go, having a span of 39 ft. 4½ in. and a length of 20 ft. 6½in., but it was the natural transition from a primary trainer to a true sailplane. The Jastraab was designed for acrobatic work, and was therefore highly stressed, with its wing mounted on the shoulder of the fuselage. A Jastraab pilot is strapped in, as are all glider pilots, with his head protected by a plastic cockpit canopy.

Empty weight of the IS-4 is 562 lb., with its loaded weight about 750 lb. Chinese students quickly took up the sport, and acrobatic meets in which Chinese IS-4 Jastraab pilots compete for prizes are very popular.

SZD-8 Jaskolka Sailplane

A mid-wing sailplane patterned after the IS-4 Jastraab, the SZD-8 Jaskolka was also a single-seat sailplane, although its span was larger and it offered greater soaring potential because of its higher lift wings. The Jaskolka entered production in Poland in 1950, and it very quickly became popular enough to become one of the most successful gliders produced by the Gliding Institute. Large numbers were produced and a variety of export models were sold to Great Britain, Belgium, Denmark, Finland, Rumania, the Soviet Union, and the People's Republic of China, as well as other Communist and non-Communist nations. In China the Jaskolka was assigned to various gliding schools and clubs, and examples were also flown by the PLAAF. The span of the Jaskolka is 52 ft. 6 in. and the length is 24 ft. 4 in. In spite of its larger size, it didn't weigh much more than the Jastraab with its empty weight of 595 lb. and loaded weight of 782 lb. Production of the SZD-8 Jaskolka ended in Poland in 1960.

SZD-9 Bocian Sailplane

The first of the true high-performance two-seat sailplanes sent to the People's Republic of China, the SZD-9 Bocian (the Polish word for "stork"), became recognized as one of the finest sailplanes in its class in the world. Prime glide ratio is 26:1. The Bocian is of all-wood construction, with a narrow-chord mid-wing spanning 59 ft. 9½ in. centered through the fuselage. The tandem cockpit placed the rear seat directly over the center of gravity to eliminate the need for ballast when the sailplane is flown as a single-seater. The Bocian is generally towed at about 90 mph by an aircraft for launching. A single-wheel landing gear is fitted with a short skid for controlled landings. The Bocian was widely exported, and by 1964 over 220 had been sold outside of Poland. Deliveries were made to Austria, Australia, Great Britain, Belgium, France, Finland, Greece, India, Indonesia, Norway, Portugal, Rumania, Syria, mainland China, and elsewhere. Sales representation has even been set up in the United States, and the Bocian has appeared at American glider meets to make it one of the few modern aircraft used in both Communist China and the United States. Loaded weight is 1,103 lb.

SZD-12 Mucha-100 Sailplane

About five years after the Polish Gliding Institute had been established, it was renamed the Szybowcowy Zaklad Doswiad Czaly, or Experimental Glider Establishment. It became commonly known as SZD, a designation used to replace the earlier IS designations of the Gliding Institute. One of the first of the new SZD products was a very attractive mid-wing single-seat sailplane called the SZD-12 Mucha-100, (Polish for "Fly Model 100"). Once again an SZD design attracted world attention, and the Mucha-100 sailplanes were sold in ten countries, including the People's Republic of China. The Mucha-100 remained in production until 1960, and later models known as the Mucha-Standard continued in production for many years afterward and are regarded to this day as among the most wanted sailplanes in the world. Mucha sailplanes are quite expensive, and with import duty they cost considerably more than most sailplanes in their class. Communist Chinese orders placed with SZD provided the Polish People's Republic with substantial trading funds in China and helped establish an active foreign trade between these two Communist nations in the mid-1950's during the Sino-Soviet "honeymoon" period.

SZD produced about 350 Mucha-100 sailplanes, and over seventy of these were exported. Red China was one of the largest customers. Span of the Mucha-100 is 49 ft. 2½ in. with a length of 22 ft. 11½ in. A fast tow, the Mucha-100 could be pulled along at 136 mph by a Yakovlev YAK-18 trainer. Empty weight is 430 lb., and loaded weight is approximately 639 lb.

Tachikawa Ki.54a Type 1 Two-Engine Advanced Trainer

Standard multi-engine trainer of the Japanese Army Air Forces during World War II, the Ki.54a was first built in 1940 and remained in production until the end of the Pacific war. The Japanese called the Model A version the 1 Sokoren, meaning Type 1 Two-Engine Advanced Trainer. Power was supplied by two Hitachi Ha.13A radial engines of 450 hp each, which gave the aircraft a top speed of 223 mph at 6,600 ft. altitude. Tachikawa produced 1,342 of these aircraft as pilot trainers; operations trainers for radio men, navigators, and bombardiers; and transport models. All of these were widely used at JAAF training schools. All of the models had a wing span of 58 ft. ¾ in., and the Model A trainer had a loaded weight of 8,996 lb. Up to eight students could be accommodated, accompanied by an instructor, but normal procedure was to take up about five students at a time. It was a stroke of good fortune for the Chinese Communists to pick up examples of such a capable trainer aircraft at the same time that they obtained multi-engined transports and bombers. The wartime U.N. code name assigned to all models of the Ki.54 series was "Hickory."

Tachikawa Ki.54c Type 1 Transport

Originally built for the Japanese Army Air Force as an advanced and operational trainer, the twin-engine low-wing monoplane airframe of the Tachikawa Ki.54 series was soon adapted for use as a light transport as the Model C for JAAF liaison and communications duties. Just about every Japanese Army Air Force unit of any consequence had a Ki.54c or two to get around. It was used like an aerial jeep. The Ki.54c had a top speed of around 230 mph and carried a crew of two plus nine passengers. Tachikawa also produced a civil version of the Model C as the Y-39. This civil version was used in Japan by Japan Air Transport and by China Airways Company on the feeder-line civil routes serving internal Japan and central and north China. When the war ended, the Chinese pilots of China Airways Company, the national airline of the government of occupied China, turned over their aircraft to the Nationalist and Communist forces, whichever first reached them to accept their surrender. As a result, both sides of the Chinese civil war flew these former Japanese transports. The Ki.54c retained the code name "Hickory" assigned to the trainer version.

Tachikawa Ki.55 Type 99 Advanced Trainer

The Tachikawa Ki.55 was the best known and most widely used advanced trainer of wartime Japan, and just about every Japanese Army Air Force pilot in the Pacific war spent some time in the 99 Koren, the Japanese nickname for the Type 99 Advance Trainer. The JAAF did much of its advanced training work close to its area of operations, and Ki.55 trainers were therefore posted to flight schools all over the empire, including the major flight school at Mukden in Manchukuo as well as a number of others on the Chinese mainland. The Ki.55 had dual controls, handled easily, and cruised at the reasonable speed of 145 mph. Use of a fixed landing gear avoided the complications of retraction and gear lowering during training, so that the student pilots could concentrate fully on flying the aircraft. The low-wing monoplane Ki.55 airframe was virtually identical to the earlier Tachikawa Ki.36 Type 98 Army Cooperation Plane, an aircraft developed for use against guerrillas in China. The Ki.55 also looked a great deal like the North American AT-6 trainer of the United States, an aircraft designed for the same training tasks. The Ki.55 trainer version of the basic Ki.36 airframe picked up the U.N. code name of "Ida" applied to the earlier aircraft. Power for the Ki.55 was a Hitachi Ha.13A 9-cylinder radial of 450 hp. Its wing span was 38 ft. 8½ in. The Chinese Communists found enough of these trainers in Manchuria to set up a number of flight schools, and many of these Ki.55 trainers served in the PLAAF well into the late 1950's. The aircraft was important enough to the Communist Chinese to preserve an example at the People's Revolution Military Museum in Peking, where it sits as a reminder of the days when the leading PLAAF aircraft were of Japanese origin.

Tchan Tia-kuo ABC-A Primary Glider

When the Szybowcowy Zaklad Doswiadczaly (or Experimental Glider Establishment) mission of gliding experts, designers, and technicians went to the People's Republic of China in the early 1950's, one of their primary objectives was to set up facilities for the licensed production of SZD gliders in Communist China at Tchan Tia-kuo, the combination gliding school and factory site that became known as the Tchan Tia-kuo Glider Manufacturing Center. Among the first models to be produced in Communist China were the IS-3 ABC-A and the SZD Salamandra 53A primary gliders, although the 53A was modified to meet Chinese needs. The factory production line was set up under the close supervision of the Polish advisers, and the student factory workers were carefully trained under their direction to be sure that they could ultimately work on their own and continue production after the Polish mission had returned to Europe. The Communist

Chinese ABC-A was virtually identical to the original Polish version, having a span of 29 ft. 6¼ in. All the material was from local sources, making the Tchan Tia-kuo ABC-A truly an indigenous Chinese product. Production of the ABC-A at Tchan Tia-kuo began in 1955, and the type was quickly allocated to flying clubs, PLA glider schools, and a number of PLAAF regiments and training establishments.

Tchan Tia-kuo SZD-8 Jaskolka Sailplane

One of the most popular sailplanes in Red China, the Tchan Tia-kuo SZD-8 is a license-produced duplicate of the Polish Jaskolka, a single-seat mid-wing sailplane with a covered cockpit. The Red Chinese have displayed rare touches of individuality with their own Jaskolkas, for many of them have been dramatically painted in a wide variety of flashy colors and markings. Naturally, red appears to be the favorite color. Quite a number entered military service and carried the PLAAF insignia on their upper left and lower right wings. The insignia carried by these gliders often varies from that found on most PLAAF combat and trainer aircraft in that the red horizontal bars and their yellow outlines are omitted, leaving only the central red star insignia. Production of the Red Chinese SZD-8 is believed to have been started in 1956. The Chinese version has an empty weight of approximately 600 lb. Because of the use of aircraft dural in this glider, separate production lines had to be set up at Tchan Tia-kuo and new manufacturing skills developed under the direction of the Polish technicians.

Tchan Tia-kuo SZD-12 Mucha-100 Sailplane

With a license to build the SZD-12 Mucha-100, the People's Republic of China had the productive capability to produce one of the most outstanding sailplanes in the world.

PLA gliding enthusiasts and air force pilots were quick to take advantage of this opportunity, and intraservice glider meets became a popular sporting event. Glider clubs flying the Chinese Mucha-100 also took part in these meets, and gliding and soaring events, such as those at Wuhan, Harbin, and Peking conducted on an annual basis, drew large crowds. The Chinese Communists even claim that sailplanes have been used in locust control in various parts of China, although the thought of handling a glider while spraying chemicals over the ground is enough to terrify even the most highly skilled glider pilot. The Chinese Mucha-100 generally cruises at just about 50 mph, and maintenance of control requires the pilot's full attention. The Tchan Tia-kuo Mucha-100 was one of the last Polish SZD designs to enter production in mainland China, and assembly work is believed to have started in the late 1950's. Span of the glider is 49 ft. 2½ in. and loaded weight is probably slightly over 640 lb.

Tchan Tia-kuo China-Salamandra Primary Glider

Although the Chinese Communists obtained a license for the single-seat SZD Salamandra 53A, when the type entered production at Tchan Tia-kuo in 1955 it was modified to a two-seat dual control trainer so that both student and instructor could fly this primary glider at the same time. The revisions in design also changed the glider's appearance; the streamlined pod cockpit fitted to the Polish version had to be removed to provide room for the second slab seat. The Tchan Tia-kuo version became known as the China-Salamandra, or Chinese Salamander. The crew of two increased the loaded weight so that it came up to about 600 lb. or more, depending on the weight of the student and gliding instructor. In Chinese service these dual primary trainers are painted a bright yellow-orange to identify the glider as a trainer and warn other aircraft to stay away. This idea was used by the Japanese, and the Chinese Communists picked it up when they obtained the Japanese trainers that were left by the defeated Japanese Army Air Force in Manchuria in 1945. Chinese gliding schools operate both the single-seat Polish model of the Salamandra and the two-seat Chinese version, providing conversion training and solo flight capability in the same type of basic glider design at the schools.

Tchan Tia-kuo Primary Glider

Gliding isn't new to China, and even under the wartime Nationalist Government there were gliding clubs scattered around the mainland. Some original Chinese designs were developed during this period, including a number of primary gliders as well as a large twelve-passenger troop and cargo glider constructed of bamboo and plywood and completed at the Institute of Aero Research at Chengtu in Szechwan province in November, 1947. These gliders were probably captured when the Communists overran the mainland, although nothing has since been heard of them. The Communists probably didn't know what to do with them and chances are that they were destroyed. The Chinese Communists have since proudly claimed the design and development of indigenous Chinese gliders, one of them a simple open-frame primary glider supplied in great numbers to clubs and schools. This may be nothing more than the reinstated production of one of the earlier Chinese designs, or a new design based on the ideas brought to China by the Polish SZD mission. The Communists did add one new idea to gliding in China. Their Tchan Tia-kuo primary gliders are launched by catapult.

Tchan Tia-kuo Jie-Fang No. 1 Sailplane

Diploma Engineer J. Niespala, head of the Polish SZD mission to the People's Republic of China, worked closely with his Chinese employers in order to establish a Chinese People's Glider Design Bureau at Tchan Tia-kuo late in 1956. The initial work was concentrated on getting the various SZD designs into production at the Chinese facility, although original design work was done on the China-Salamandra to convert it to dual control. The prime goal of the Glider Design Bureau at Tchan Tia-kuo was still the creation of indigenous Chinese Communist sailplane designs. As soon as the hammer-and-tong work of getting the office organized and into operation was completed, Chief Designer Niespala assembled a design team to start on their own designs and initiate work on their first project. This was to become the first original high-performance sailplane design developed in the People's Republic of China. Although the design was under the direction of Dr. Niespala, Chinese engineers did the bulk of the work. The prototype was completed early in 1958 and made its first flight on May 10. It was called the Jie-Fang No. 1, translating as Liberation No. 1. In appearance it is a clean-looking mid-wing monoplane with a glazed canopy over a two-seat tandem cockpit layout. It looked very much like a SZD-9 Bocian sailplane, a type already being imported from Poland, and there is little question that the Polish glider contributed to its design. The Jie-Fang No. 1 entered production at Tchan Tia-kuo for use by the PLAAF and various glider clubs and flight schools.

Tupolev SB-2 Heavy Bomber

The Soviets' delivery of modern twin-engined Tupolev SB-2 heavy bombers to the Red Chinese at Yenan in 1937 and 1938 has been obliquely referred to in the writings of the Chinese Communists, but the actual fact has been shrouded in secrecy and awaits positive confirmation. The revelation that Polikarpov R-5 general-purpose biplanes were actually delivered and received by the Chinese Communist Army in 1937 indicates that such aircraft deliveries were possible, and lends credence to the SB-2 delivery reports. This advanced mid-wing monoplane bomber was first produced in 1936. It had a crew of three and a top speed of 236 mph, making it faster than many of the contemporary fighter aircraft of the period. A substantial number were supplied by the Soviet Union to the Chinese Nationalists for use against the Japanese at a time when the Soviet Union's stated policy was to support the Nationalists in their fight against "Fascist" Japan. Deliveries to the Nationalists started late in 1937 and were terminated in 1939, when the Russians altered their planning to concentrate on their problems in Europe and the fighting in Finland. Use of this aircraft by the Chinese Communists would no doubt have required Russian crews, and it is known that volunteer Russian pilots did serve with the Chinese Reds in the early days of the Sino-Japanese fighting. Russian volunteer crews also served with the SB-2 bombers assigned to the Nationalists, and the SB-2 was the most numerically important Chinese Nationalist bomber until the arrival of American aircraft in the later years of World War II. Power for the SB-2 was two Russian M-34 liquid-cooled inline engines of 830 hp each.

Tupolev TU-2 Attack Bomber

Standard Russian attack bomber when the war in Europe ended in 1945, the Tupolev TU-2 remained in full-scale production until 1948 to meet the continuing needs of the Soviets and the many satellite nations that received the type for the new Communist air forces being created in Europe and Asia. The TU-2 first made its appearance in 1943 after years of design and production problems created by wartime conditions. The final result led to a coveted Stalin Prize—a Soviet government award for excellence that included a substantial amount of money—for its designer Andrei Nikolaievich Tupolev. The TU-2 was a large aircraft as ground-attack types go, and it had a wing span of 61 ft. While it normally carried a ton of bombs, it could be overloaded to carry over 6,000 lb. of arms, making it a formidable foe for use against any ground army. It is estimated that over 300 were supplied to the People's Republic of China by the Soviets, together with supporting Petlyakov PE-2 trainers. Power for the TU-2 was supplied by two Shvetsov ASh-82FNV 14-cylinder radials of 1,850 hp each, the same engines used by the companion Lavochkin LA-9 and LA-11 fighters also supplied to Communist China; this simplified Chinese maintenance problems. Top speed was 340 mph. NATO code name for the TU-2 is "Bat."

Tupolev TU-4 Heavy Bomber

One of the most shocking sights to an American in the middle 1950's would have been the observation of what appeared to be American B-29 bombers over mainland China escorted by North American P-15D Mustang fighters in training exercises, with all of these aircraft in Communist Chinese insignia. This would somehow look like World War II in reverse, and yet it happened. When the Russians refused to return three American Boeing B-29 Superfortress bombers that had landed intact at Vladivostok in the Soviet Union in 1944 after raids on Japan, the American War Department quickly knew what it was up against. The B-29 was the most advanced strategic-bombing weapons system in the world; literally all of the hard-earned American experience in this phase of air weaponry was packed into this single highly sophisticated machine. The Soviets got the whole package in a single windfall, with enough machines to test, study, and dismantle at the same time. Under direct orders from Stalin the noted Russian designer Andrei Nikolaievich Tupolev was given the personal responsibility of recreating the B-29 for Russia in a move designed to completely negate the American advantage in this field in the event of any postwar showdown. Only the possession of the Atomic Bomb gave the United States a sure advantage at the time, and even that lasted only for a short period. Tupolev's design group re-engineered the entire aircraft, including the complex electrically operated fire control system, and by March, 1945, the Russian version was already in production as the Tupolev TU-4. It earned Tupolev another rewarding Stalin Prize. The powerful 2,200 hp 18-cylinder Wright Cyclone engines of the B-29 were also copied and further developed as the Shvetsov ASh-73TK to give the TU-4 virtually the same performance as its American counterpart. About 1,200 of these aircraft were produced, and about 400 are said to have been delivered to the Chinese Communists. The few remaining in use in mainland China suggest that the reported delivery figures were exaggerated, although without question the inexperienced Chinese Communists wrecked a substantial number of these aircraft in training and operational flights, making the entire effort a costly experience. The fact that the Russians could recreate the TU-4 from the B-29 in less than a year is a fantastic engineering achievement, and it is said that the Russian engineers literally worked night and day until they had not only redesigned and redrawn every element of the aircraft but had also achieved an understanding of every working part. The NATO code name for the TU-4 is "Bull." Perhaps they should have called it "Brain."

Tupolev TU 70 Paratroop Transport

It is sometimes startling to see how designers in two different countries solve the same problems, and then to compare their solutions. The Boeing Airplane Company in the United States and the Tupolev Design Bureau, working in the Soviet Union, had the same input to work with—the airframe of the Boeing B-29 Superfortress and its Russian equivalent, the

Tupolev TU-4 heavy bomber. Both nations pursued development of a pressurized military passenger and paratroop transport version. Boeing's answer was a completely new dual-decked fuselage-mounted midway on a set of four-engined B-29 wings and tail assembly providing a transport that could carry 134 armed troops at a top speed of 350 mph with a loaded weight of 135,000 lb. in an aircraft spanning 141 ft. 3 in. They called it the XC-97. Power was four 18-cylinder Wright R-3350-23 Cyclones of 2,200 hp each. The Russian group at the Tupolev Design Bureau came up with a new fuselage, but they dropped the wing to the more conventional low-wing position. Their answer carried 110 armed paratroops and a crew of four or five in a Douglas-type fuselage at a speed of 298 mph in an aircraft spanning 141 ft. Power was four Shvetsov ASh-82FNV 14-cylinder radials of 1,850 hp each. Clearly the American approach had the edge, but it also had considerably more power. Tupolev's design was designated the TU-70, or more probably the TU-7 Model O, and NATO gave it the code name "Cart." A few were reportedly delivered to Communist China. A civil version was also produced to carry seventy-two passengers, but it didn't seem to catch on.

Tupolev TU-14 Patrol Bomber

A contemporary of the Ilyushin IL-28 twin-jet bomber, the Tupolev TU-14 light attack and patrol bomber looks somewhat like the Ilyushin type, although it is considerably longer and lacks the swept tail surfaces. First flown in prototype form in 1949, the Tupolev TU-14 was produced for the Soviet Navy in quantity and began to enter service in 1951. It made its first public appearance over Moscow on Aviation Day at Tushino in 1951, the year after the introduction of the IL-28. The TU-14 served primarily as a land-based torpedo bomber, and it could carry two internally mounted 1,850 lb. naval torpedoes in its bays. Bombs, anti-shipping mines, or depth charges could be carried in place of the torpedoes. Russian naval patrol units based on the Baltic Sea, on the Arctic Sea, and in the Far East flew the TU-14 throughout the 1950's and the early 1960's. Between fifty and sixty were supplied to the PLANAF for shore patrol duties, and they were later joined by torpedo-carrying versions of the IL-28, the latter modified by the Chinese Communists. The TU-14 was long and skinny, having a wing span of 83 ft. 8 in. and a length of 80 ft. 6 in. Power was provided by two Klimov VK-2 turbojets of 5,500 lb. thrust, each in squared-off nacelles similar to those on the IL-28. Later models had the more powerful Klimov VK-2R of 6,600 lb. thrust. The crew of three to four included a tail gunner's position under the large vertical tail. Loaded weight was about 50,000 lb. The small number supplied to Communist China suggests that these aircraft were phased out of service some time ago. The NATO code name is "Bosun."

Tupolev TU-16 Medium Bomber

Standard Soviet medium bomber in the late 1950's and early 1960's, the Tupolev TU-16 became the line-leader of Soviet influence in nations not quite in the Soviet orbit but perhaps friendly toward Communism. In the early 1960's TU-16's were sent to Egypt, where they were prime targets of the Israeli Air Force in the six-day war in June, 1967; to Indonesia, where they would have been a dagger at the throat of British and American bases in the Far East if the pro-Communist coup had succeeded; and possibly, but only possibly, to Communist China. In the spring of 1962 reliable sources reported that the Soviet Union had sold a small number of swept-wing TU-16 bombers to Communist China as replacements for the aging PLAAF IL-28 bombers. The reported sale aroused India at a time when China and India were bound to strong opposition views in regard to the Sino-Indian border.

This was also at a time when the Soviet Union was courting India's friendship. Reports of the sale of TU-16 bombers to Communist China quickly disappeared, and it is possible that an anticipated sale was terminated at the time. The possibility exists that at least one or two of the TU-16 bombers reached Communist China, just as prototypes of the MiG-21 did somewhat earlier, and provided the National Aircraft Factory at Shenyang with something to study and possibly duplicate. In Russian service the TU-16 replaced the piston-engine TU-4 bomber. Introduction of the TU-16 in 1954 created quite a stir in American circles when it was realized that a swept-wing bomber as large as this was being powered by only two turbojets, suggesting Russian engines of much greater power than American ones of the period. Study of the Tupolev TU-104 jet civil transport, utilizing the TU-16 wing and tail surfaces, gave the United States its first accurate data on the TU-16. The NATO code name "Badger" has been assigned to the TU-16; it is one of the best known of the NATO code names because of the many TU-16 bombers flying around the world. Power is supplied by two massive Mikulin AM-3M turbojets of 19,180 lb. thrust each. Span is 100 ft., length 121 ft., loaded weight 150,000 lb., and top speed 620 mph.

VEB Flugzeugwerke Dresden IL-14P Transport

When the German Democratic Republic started a crash program to create a modern aircraft industry in the spring of 1955, it was developed under the capable direction of two former Junkers engineers, both of whom had worked in Russia following the defeat of Germany in 1945. Dipl. Engineer Fritz Freytag became the Technical Director and Design Chief of the Dresden Aircraft Plant of the Volkseigener Betriebe, translated as Publicly Owned Factories, and Dipl. Engineer Brunolf Baade was put in charge of production at the same plant. VEB-Dresden was responsible for the development and production of transport aircraft, and the first type to be produced was a license-built East German version of the Russian Ilyushin IL-14. Considerable redesign work was undertaken by Freytag, creating an IL-14 model that was lighter in weight than the Russian version, as well as slightly smaller in span and length. East German refinements were also added to the Shvetsov ASh-82T engines, and the German version was produced at the VEB-Karl Marx Aircraft Engine Plant as the ASh-82T-7. Airframe production began in 1956 as the VEB-Dresden IL-14P, and about eighty were built before production ended in 1959. IL-14P transports were sold to the national airlines of North Viet-Nam, East Germany, Rumania, Bulgaria, Hungary, and Communist China. The Chinese were so pleased with their original shipment that they ordered more IL-14P's in the summer of 1959. Deliveries for this second order started in the fall of the year by means of direct flights from East Germany to Peking, China, with refueling stops being made across the Soviet Union. The Red Chinese IL-14P was fitted to carry a crew of five and either eighteen or twenty-six passengers. Wing span was 103 ft. 4½ in. and the length 67 ft. 4½ in. In the twenty-six-passenger configuration, loaded weight was 35,350 lb. and the top speed 245 mph. The production success of the IL-14P was not enough to satisfy Dr. Freytag, for his dream was to create the most modern jet transport in the Communist world. He almost accomplished this with the VEB-Dresden Type 152, a four-jet swept-wing transport that looked very much like a civil version of the contemporary American B-47 Stratojet, a jet type then being flown by the Strategic Air Command of the USAF. The construction of twenty Type 152 transports was under consideration when the whole program received a mortal blow with the loss of the first prototype in a pilot error crash during its second flight on March 4, 1959. It wasn't until August 26, 1960, that the second prototype made its first test flight, and by this time VEB-Dresden IL-14P production had come to an end and there were rumors that the whole VEB complex would cease to produce aircraft and be converted to industrial production. Less than two months later, Dr. Freytag fled to the West. Brunolf Baade remained behind, and the following year found himself producing potato planters, industrial engines, and prefabricated windows in the same facilities as that used for the production of the IL-14P and planned for the Type 152. Had the advanced Type 152 reached production, it is all but certain that Communist China would have been a major customer.

Vickers 843 Viscount Medium Transport

One of the most fabulously successful commercial aircraft of the 1950's, the British Vickers Viscount series of four-engine turboprop transports first entered service in 1953. By 1960 over 250 of the original 700 series were in service in over twenty countries, and the Viscount was one of the best-known transports in the world. It came into service at a most opportune time, for the British aircraft industry was still smarting over the disastrous showing made by the early models of the De Havilland Comet, the world's first production jet transport. In 1956 the original Viscount series was "stretched" to create the model 800, with various models fitted out to carry from forty-four to seventy passengers. Then, in 1958, more powerful models of the Rolls Royce Dart turboprop used by the Viscount became available as the Mark 525 and Mark 541, the latter rated at 2,350 hp each, and by the spring of 1960 over seventy of the 800 series had been sold to airlines in a dozen countries. The Chinese Communists began negotiating with the British for the purchase of Viscounts as early as January, 1961, and in spite of strong protests by the United States, an order for six Viscount 843 transports was concluded by the end of the year. The order was an important one for the British, for it suggested a growing trade with Communist China. The sale price was $930,000 for each aircraft, with spares and supplies running the total up to almost $10 million. The first Communist Chinese Viscount arrived at Hong Kong on July 6, 1963. A week later it left for the People's Republic in the hands of its Communist Chinese CAAC crew. The second and third examples followed in August and September. The normal production procedure of painting customer markings on the CAAC Viscounts and flight crew take-over at the factory was not followed in this case. The Viscounts were flown by British crews to Hong Kong completely devoid of markings, with the exception of British civil registrations. This was considered necessary in order to avoid political complications and demonstrations en route, particularly in India. As the 843's arrived in Communist China, they immediately entered service, becoming CAAC workhorses. They received even-number civil registrations from 402 to 412, a practice followed on CAAC transports. The Red Chinese Viscounts carry a crew of three and up to seventy-three coach-class passengers. Wing span is 93 ft. 8½ in., and loaded weight up to 72,500 lb. They probably should have a code name, but nobody ever got around to it. Cruising speeds are just short of 400 mph.

Vought V-65-C1 Lenin General Purpose

The delivery of thirty-eight Vought V-65-C1 Corsair light bomber two-seat biplanes to the Nationalists in 1932, coupled with the arrival of the new Curtiss Hawk II Fighters, gave Chiang Kai-shek complete control of the air over China for the application of raw power against the Communists and as a defense against Japan. In 1932, an unofficial American air mission under the imaginative and efficient direction of then Colonel John H. Jouett was sent to China to help create a modern Chinese air force. This led to the evaluation and purchase of these modern types by the Nationalists. The Vought V-65-C1 flew at 180 mph, weighed in at 3,256 lb. loaded for combat operations, and was powered by a 600 hp Pratt and Whitney Hornet. In China, it was known as the New Corsair, to distinguish it from examples of the earlier export version of the Vought O3U-1 Old Corsair already in Chinese service. To the Chinese Communists the New Corsair was one of the most hated aircraft in the Nationalist inventory, for it was used to bomb the Communists along most of the route of the Long March, in company with the Douglas O-2MC4 type also in Nationalist service. Mao Tse-tung's second wife, making the march while pregnant, was severely wounded in one of the raids, and chances are that one of the Vought V-65-C1's was responsible. The Chinese Communists never had much of a chance to use their captured V-65-C1, which they named Lenin, for it was soon lost to them after being shot down by a Nationalist aircraft on October 5, 1936. The Chinese Communists never forgot that the aircraft that were the most dangerous to them in the middle 1930's were of American origin.

Watanabe E13A1 Type O
Three-Seat Reconnaissance Seaplane

As the Japanese Navy's most widely used reconnaissance floatplane of World War II, over 1,400 of these single-engine twin-float monoplanes were produced between 1940, the year it was first accepted for production, and June, 1945, when construction was terminated at the Kyushu Airplane Company, Ltd. The E13A1 design was started by the Aichi Aircraft Company, Ltd., in 1937, to replace the old biplane reconnaissance floatplanes then being used by the Imperial Japanese Navy. The prototype of the E13A1 first flew in 1938 and immediately found itself in competition for a lucrative Navy contract with an aircraft of similar design created by the Kawanishi Aircraft Company, Ltd., a rival producer. This highly involved process serves to demonstrate the extreme competitiveness of Japanese business in the years prior to Pearl Harbor and provides a clear picture of Japan's abhorrence of the threat of Communism in the 1930's and 1940's. The original Aichi design was finally selected after a lengthy series of design changes and received the Navy's Type O production designation, a number based on the last digit of the current Japanese year, which was then 2600 (corresponding to the year 1940 in Western calendars). It wasn't until 1941 that production got under way at Aichi, but work was soon reassigned to the Navy's Hiro Arsenal and the Watanabe Ironworks to enable Aichi to concentrate on a series of carrier dive bombers just accepted by the Imperial Navy. Watanabe produced most of the E13 series of aircraft, and it became generally known as a Watanabe, and later a Kyushu, type. The E13A1 operated in China and across the Pacific as a reconnaissance plane and light bomber, and was even used as a dive bomber. It was catapult-launched from cruisers or battleships and often operated from sheltered island bases across the southeast Pacific during the Pacific war years. Its last major wartime use was as an antisubmarine patrol plane carrying magnetic submarine-detection gear. In this guise it operated out of bases located along the coast of mainland China in order to protect the Japanese shipping lanes in 1944 and 1945, with many remaining at their bases at the war's end to be picked up by the Nationalists and the Chinese Communists. Power for the E13A1 was provided by a 14-cylinder Mitsubishi Kinsei 43 radial of 1,000 hp, which gave it a top speed of 239 mph. The E13A1 received the wartime Pacific code name of "Jake."

Yakovlev YAK-9P Fighter

Best known of Russia's World War II fighters, and produced in greater numbers than any other Russian piston engine fighter types, the YAK series served throughout the war in

YAK-1, YAK-7, and YAK-9 models. Though the original examples were of wood and metal construction, in 1944 the conversion to all-metal construction was accomplished with the YAK-9U model. The final production model of this famous series was the YAK-9P, which entered production soon after the war was over. The YAK-9P quickly became Russia's major export fighter and it soon showed up in just about every country receiving military aid from the Soviet Union in the late 1940's and the early 1950's. It became a standard fighter of the PLAAF in 1950, just prior to the Korean War. Power was from a 12-cylinder Klimov M-107A of 1650 hp, which gave the fighter a top speed of 415 mph. Its span of 32 ft. 9¾ in., length of 28 ft. 6½ in., and loaded weight of 6,985 lb. made it considerably smaller and lighter than the North American P-51D Mustang, an aircraft with which it was often compared. When NATO code names first came along, the YAK-9P received the name "Frank," the same name that had been used earlier for the World War II Nakajima Ki.84 Type 4 fighter of the Japanese Army. A YAK-9P of the North Korean Air Force was the first Communist aircraft to fly over the 38th parallel in June, 1950, dramatically announcing the start of the Korean War.

Yakovlev YAK-11 Advanced Trainer

Just as the Japanese used the airframe of the Nakajima Ki.27 Type 97 fighter for their Manshu Ki.79 trainer, the Russians adapted the airframe of their leading wartime Yakovlev YAK-9 fighter soon after the end of World War II to a two-seat trainer aircraft. By using existing tools, jigs, and production facilities, the Soviets were able to come up quickly with a dependable trainer type that handled well and had all of its basic design bugs removed before it even flew for the first time. The YAK-11 proved to be even more durable than its parent fighter, for modern versions are still in daily use in the

Soviet Union and other Communist nations well over twenty years after its creation. The YAK-11 formed the training backbone of the many post–World War II air forces created with help from the Soviet Union in Europe, the Middle East, and Asia. YAK-11 trainers have flown with the air forces of Egypt, Syria, Afghanistan, the People's Republic of China, Albania, Hungary, Austria, Poland, Cuba, Rumania, Czechoslovakia, and virtually anywhere that Russian influence has been felt over the past two decades. A 680 hp Shvetsov ASh-21 radial gives the YAK-11 a top speed of 286 mph, and often a single 7.7 mm machine gun is carried on the left side of the cowling for firing practice. The YAK-11 has been code-named "Moose" by NATO.

Yakovlev YAK-12 Communications

Looking for all the world like a typical American light plane in the high-wing Piper or Cessna class, the YAK-12 was in fact designed by the dean of the World War II fighter aircraft designers, Alexander S. Yakovlev. Although it might seem strange that a noted designer of the YAK series of fighters should tackle the design of a light aircraft, it was perfectly normal for Yakovlev, for he had long been interested in gliders and light communications and sport aircraft. The YAK-12 grew out of a World War II requirement to replace the obviously old and obsolete Polikarpov PO-2 utility biplane, although the PO-2 went on to serve for years after the war. The YAK-12 was designed to make use of the same Mikulin M-11FR 5-cylinder radial engine of 160 hp that was used by the later models of the PO-2, and the first production models began to enter Soviet service in 1946 and 1947, described as the YAK-12 Aeromobile. The initial YAK-12 was slightly larger and heavier than the American Cessna 170 of the same period; the Russian aircraft had a span of 39 ft. 6 in. and a loaded weight of 2,650 lb. carrying a pilot and three passengers. The original model, as first supplied to the People's Republic of China, had a wooden wing and a steel tube fuselage and was powered by a Mikulin M-11D engine of 145 hp. This version was first given the NATO code name "Crow." Later models had the new 260 hp Ivchenko AI-14R radial, and the YAK-12R version was introduced in 1952 and assigned the NATO code name "Creek-B," reflecting a change in the NATO codes. An all-metal model entered production in 1955 as the YAK-12M and was known as the "Creek-C," and a tapered-wing model came out in 1957 as the YAK-12A, coded by NATO as the "Creek-D."

Yakovlev YAK-14 Transport Glider

Long-range infantry assault by means of troop-carrying gliders was one of the new military concepts to come out of World War II. This technique first came into prominence with the use of gliders in the German invasion of Crete in May, 1941, and soon every major world power was adding some form of troop-carrying glider capability to its ground and air forces. The British and American invasion of France and the Low Countries in 1944 included a substantial glider force utilizing massive troop-carrying gliders that were dropped in swarms over the countryside. The Soviet Union noted these lessons well, and in 1944 work started on a number of large troop and cargo gliders for military use. The largest of these was a heavy all-wooden cargo glider which ultimately appeared publicly for the first time at the Tushino air display on Soviet Aviation Day in 1949. Designated the YAK-14, the craft had a large wing spanning 85 ft. 10¼ in. mounted on the shoulder of a bulky fuselage 60 ft. 6 in. long. It is this type that is believed to have been supplied to the People's Republic of China for its assault forces. NATO code name for the YAK-14 is "Mare." In operational use a single YAK-14 is towed by an Ilyushin IL-12 or IL-14 transport at a towing speed of up to 186 mph. The Chinese Communists may also have been supplied with a similar but smaller troop glider known as the Zibin KZ-20, a design created by Senior Soviet Glider Engineers P. V. Zibin and D. N. Kolesnikov. The KZ-20 entered production in 1944. NATO code-named "Mist," the KZ-20 has a span of 80 ft., is 53 ft. long, and has a tow speed of 144 mph. Gliders remained important until the advent of the large troop-carrying helicopter that eventually took over the troop assault role.

Yakovlev YAK-16 Transport and Advanced Trainer

Built for the same variety of uses as the Japanese Tachikawa Ki.54 series and the Russian replacement for this type in PLAAF use, the Yakovlev YAK-16 was a clean-looking twin-engined low-wing monoplane of conventional design. First flown in 1946, and somewhat larger than its earlier Japanese counterpart, the YAK-16 was developed as the multiengine companion to the YAK-11 trainer. It provided the Soviet Union with a twin-engined trainer in this class some years after most other nations. The flurry of post–World War II Russian designs made it look as if the Soviet Union was trying to catch up with the rest of the world in a rush in order to establish an inventory of aircraft types for all uses. Perhaps many of the postwar types had already been designed during the years of fighting and were held back in order to maintain production of as many combat types as possible for the Great Patriotic War, as the Russians called World War II. The YAK-16 made its debut as a light transport for eight to ten passengers, while its major military function was that of a multipurpose trainer in a wide variety of configurations. It was used as a multiengine crew trainer, navigational trainer, bombardier trainer, cargo transport, VIP liaison and communications type, as well as a glider tug. Appearances varied between solid nose and glazed nose sections, depending on the duties, and a navigational version had a row of sighting astrodomes on top of the fuselage. The Russians tried valiantly to sell the YAK-16 as an export type, both within and outside the Soviet bloc, exhibiting the civil transport version at the International Trade Fair at Poznan, Poland, in April, 1948, and later in the year at Budapest, Hungary. This was followed by trade shows in Finland, Czechoslovakia, and Rumania. At that time the Soviets tended to rate their labor costs much too highly, and the YAK-16 was priced out of the competitive market. It would appear that only in Communist China, where costs were ignored in the interest of creating a new air arm, did the YAK-16 take root on foreign soil. This open-end approach later plagued the Chinese Communists when they received the bill from Premier Khrushchev's government for the generous Russian aid of the early 1950's. A few YAK-16 transports saw civil service on Aeroflot, Russia's national airline, and the type was widely used in the Soviet Air Force in its trainer roles. Power was from two 7-cylinder Shvetsov ASh-21 radials of 680 hp each. Span was 65 ft. 7 in., length 49 ft. 6 in., and cruising speed 173 mph, with a top speed of 190 mph. Loaded weight in its transport version was 14,112 lb. The NATO code name for the YAK-16 is "Cork."

Yakovlev YAK-17UT1 Advanced Trainer

When the era of jet aviation first came to the Soviet Union at the close of World War II, the Russians were quick to adapt the new method of power to their military aircraft. The first of the new jet fighters to enter service in the Soviet Air Force was the Yakovlev YAK-15, mounting two 20 mm cannon and flying in prototype form in April, 1946, less than a year after the end of the war in Europe, and only seven months after the war had ended in the Far East. This rapid technological gain was made possible by the simple expedient of taking a YAK-9U fighter airframe and replacing the piston engine with a captured German Junkers Jumo-004B axial-flow turbojet of 1,980 lb. thrust. The German jet engine was slung under the rebuilt nose section to create a mid-wing jet out of the original low-wing fighter. Production models utilized the RD-10 turbojet of 1,890 lb. thrust, a Russian adaptation of the original German jet engine, giving the YAK-15 a maximum speed of almost 500 mph, considerably higher than the most advanced YAK-9P piston engine fighter with the same basic airframe. Within a year the YAK-15 was completely redesigned as the YAK-17 fighter to square off the wings and tailplanes for ease of production, strengthen the airframe for the greater stresses of jet power, and incorporate a completely redesigned landing gear with a nosewheel to keep the jet exhaust above the surface of the ground and avoid damage to the airfields from which the new fighter would be flying. This version entered service in 1948 and received the NATO code name "Feather." It was supplied in numbers to the Communist satellite air forces in Eastern Europe. Soon thereafter a two-seat jet fighter conversion trainer version appeared as the YAK-17UTI, a number of which were supplied to Communist China. Top speed is reported to be around 480 mph. This trainer version has received the NATO code name "Magnet."

Yakovlev YAK-18 Primary Trainer

The Russians waited until the end of World War II before they made the noncombatant and training aircraft equipment changes that were long overdue. When these changes finally took place it seemed that the Yakovlev Design Bureau got most of the jobs. After the YAK-12 started to take over some of the duties of the Polikarpov PO-2 in the late 1940's, about the only other truly ancient trainer type that remained in service was the prewar Yakovlev UT-2 primary trainer. This was a 1935 two-seat monoplane design with open cockpits and a fixed landing gear powered by the venerable Mikulin M-11 radial engine. Its replacement was a modern adaptation of the same basic design with a retractable landing gear and a covered canopy for its crew of two. Some of the design work on the original YAK-12 showed up on the new trainer, designated as the YAK-18, particularly the helmeted cowling of the 5-cylinder Mikulin M-11FR radial engine of 160 hp. Designed in 1944, by 1946 the YAK-18 began to enter service as the new standard primary trainer of the Soviet Air Force. The rest of the world had a chance to get a good look at the YAK-18 at the International Trade Fair in Poland in the spring of 1948. The YAK-18 soon became one of the most widely exported aircraft produced in the Soviet Union, because it met the training needs of all of the new postwar air forces formed under Russian direction in Eastern Europe and the Far East. It looks as if the YAK-18 will far exceed the lifetime record of its earlier UT-2 parent aircraft. After over twenty years of service the YAK-18 is still being produced in the Soviet Union, although the models are highly modified. A tricycle-landing-gear version was built as the YAK-18U, and a cleaned-up version with a 260 hp Ivchenko AI-14R radial engine entered production in 1957 as the YAK-18A standard piston-engine primary trainer. It now serves the Soviet Air Force in this capacity. A single-seat version of the YAK-18A is also produced as the YAK-18P for aerobatic and sport flying, winning a number of international events in this field. The versions delivered to Communist China were of the original YAK-18 model, having a span of 33 ft. 9½ in., a length of 26 ft. 6 in., and a loaded weight of 2,470 lb. The Russians still haven't given up on further development of the YAK-18 series. At the Paris Air Show in June, 1967, they introduced a new four-place utility transport version as the YAK-18T, and this doesn't appear to be the end of the line. Most models go under the NATO code name "Max."

SELECTED BIBLIOGRAPHY

All sources and publications listed are in English unless otherwise noted.

Reports, Original Documents, and Reference Works

Annuaire de L'Aéronautique (Aeronautical Annual). Paris, 1927. French aircraft yearbook. 1927 edition includes data on aircraft exported to China in the 1920's.

China: The Roots of Madness. T. H. White Television Production. January 31, 1967. Assembled and narrated by Theodore H. White. Numerous film clips of Nationalist and Red Chinese aircraft. Refers to Mao Tse-tung's first flight in the American transport given to the Chinese Communists.

CNAC–Listing of Impounded Chinese Aircraft. Crown Colony of Hong Kong: Secretariat General, 1950. Unpublished document. Listing of China National Aviation Corporation civil aircraft impounded in the Crown Colony of Hong Kong that were briefly turned over to the Chinese Communists. In Chinese.

China and U.S. Far East Policy. Washington, D.C.: Congressional Quarterly Service, April, 1967. A factual account of relations between the United States and Nationalist and Communist China from 1945 to early 1967. Includes data on the aerial encounters of the PLAAF.

China Handbook. Vol. VII (1953–54). *China Yearbook*. Vols. VIII (1954–55)–IXX (1965–66). Taipei, Taiwan: China Publishing Co. Official government records of the Republic of China. Contain details on annual national progress. Extensive notes on Communist

China, including coverage of Communist Chinese air power and aerial engagements between the PLAAF and the Nationalist Chinese Air Force.

China Yearbook. Vol. XVIII (1936). Shanghai: North China Daily News & Herald. Annual Chinese records containing information on the development of military and civil aviation in China up to 1936. The 1936 edition covers the Nationalists, various war lords, and the threats posed by Japan and Communism.

Documents on Communism, Nationalism, and Soviet Advisers in China, 1918–1927. New York: Columbia University Press, 1956. Translations of the papers seized in April, 1927, during the raid on the Soviet embassy in Peking under the leadership of General Chang Tso-lin of Manchuria. Fascinating reading. Original material on the development of Republican Chinese, Communist, and KMT aviation under Russian guidance in the 1920's. Translated from Chinese and Russian.

Foreign Relations of the United States, Far East. Vols. III, IV (1938); III, IV (1939); IV (1940); IV, V (1941); I (1942); III, V (1943); V (1944). Washington, D.C.: Government Printing Office, 1954–65. Original documents on all phases of American contact with China. Many documents on Chinese air power, Communist requests for aid, military aviation, and American aid.

Handbook on the Chinese Communist Army. Pamphlet No. 30–51. Washington, D.C.: Department of the Army, September, 1952. A complete handbook on the military equipment of the Communist Chinese ground, naval, and air forces. Details on the PLAAF during the conquest of the mainland and early in the Korean War. One of the few sources to mention Soviet aeronautical aid in the mid-1930's.

Jane's All the World Aircraft, Vols. XXI (1929)–LVIII (1966–67). London: Sampson Low, Marston. International aviation yearbook of all major aircraft built during the current year. Good coverage of Red Chinese aircraft in the late 1950's and early 1960's.

Konstrukcje Lotnicze Polski Ludowej (Aircraft Constructions of the Polish People's Republic). Warsaw: Wydawnictwa Komunikacji i Lacznosci, 1965. A detailed description of all Polish aircraft since World War II. Extensive notes on the SZD gliders exported and licensed for production in Red China.

Leaders of Twentieth Century China. Hoover Institute and Library Series IV. Stanford, Calif.: Stanford University Press, 1956. Includes references to personalities influential in Chinese aviation.

Mainland China Organizations of Higher Learning in Science and Technology and Their Publications. Washington, D.C.: Government Printing Office, 1961. A guide to selected Red Chinese universities, colleges, institutions, and aeronautical schools.

The Military Balance, 1966–1967 and 1967–1968. London: Institute for Strategic Studies, 1966 and 1967. Annual estimates of the world's major military forces, including those of the People's Republic of China. Lists Red Chinese aircraft strength and estimates current numbers of various aircraft types.

PAULEY, EDWIN W. *Report on Japanese Assets in Manchuria to the President of the United States, July 1946.* Washington, D.C.: Government Printing Office, 1950. Extensive report. Illustrations of the industrial base of Manchuria, and notes on the removal of facilities by Russians. The Manshu Airplane Manufacturing Company is discussed in detail.

Private Code–The Far East Aviation Co., Ltd.; The Far East Flying Training School, Ltd.; Far East Motors. (3d ed.). Hong Kong: The Far East Aviation Co., December, 1933. List of products, sales, suppliers, customers, civil and military airports, governmental and war lord contacts, and provincial aviation headquarters in Nationalist and Communist China.

Programs of Japan in Indo-China. Assemblage 56. FCC intercepts of short wave broadcasts from Radio Tokyo and affiliated stations from December, 1941, to May 24, 1945, and from OSS sources. Honolulu, Hawaii: Office of Strategic Services (OSS), Research and Analysis Branch, August 10, 1945. Unpublished. Brief comments on the air forces of Manchoukuo and the wartime National Government of China at Nanking. Original material on occupied French Indochina and the formation of independent Cambodia, Laos, and Vietnam after May 10, 1945.

Sekai Koku Nenkan-Showa 10 (World's Aircraft Yearbook). Tokyo: Umi-to-Sora, 1935. A pictorial description of major world aircraft types in production and use in 1935. Includes Chinese designs not described in Western sources.

U.S. Army Area Handbook for Nepal. Washington, D.C.: Government Printing Office, May, 1964. Guide to Nepal. Includes data on the use of Communist Chinese aircraft in Nepal.

Who's Who in Communist China. Hong Kong: Union Research Institute, 1966. Includes key personnel in the Red Chinese aircraft industry and the PLAAF.

World Survey of Civil Aviation, Southwest Asia. Washington, D.C.: Government Printing Office, 1961. Inventory of civil aircraft in the nations bordering Red China.

YANG, N. C. *Postage Stamp Catalog of People's Republic of China.* Kowloon, Hong Kong: Yang's Stamp Service, 1964. Listing of Red Chinese stamps not catalogued in the West. Airmails, aviation commemoratives, and Korean War issues on aircraft types and aeronautical events.

ZIRKLE, HELEN K. *Postage Stamps and Commemorative Cancellations of Manchoukuo.* New York: Collectors Club, 1964. Listing of Manchoukuoan stamps not catalogued in other sources. Airmail and aviation commemoratives on aeronautical events and aircraft types later used by Red China.

BOOKS, ARTICLES, AND PAMPHLETS

ANDERSON, JACK. "What Really Happened in Korea," *Parade,* September 22, 1963. Discussion of China's entry into the Korean War and of the tacit agreement between Red China and the United States whereby the PLAAF would not bomb South Korea. Based on unreleased documents and interviews.

BARMINE, ALEXANDER. *One Who Survived.* New York: G. P. Putnam's Sons, 1945. Memoirs of a Russian defector. References to Soviet aircraft and personnel sent to China in the 1930's and early 1940's.

BARNETT, A. DOAK. *Communist China: The Early Years, 1949–55.* New York: Frederick A. Praeger, 1964. Economic and political development in Red China. Extensive notes on Red China's financing of, and participation in, the Korean War, including aircraft purchases and use of air power.

BEECHER, WILLIAM. "Gauging China's Will," *The Wall Street Journal* (March 5, 1965). Report on the confrontation between the United States and Red China along the Chinese border. Reference to Red China's use of the MIG-21 aircraft.

BRUCE, J. M., AND JEAN NOEL. *The Breguet 14.* London: Profile Publications, 1967. A history of the Breguet 14 aircraft.

CAIN, CHARLES W. AND DENYS J. VOADEN. *Military Aircraft of the U.S.S.R.* London: Jenkins, 1952. Pictorial and factual survey of current Russian aircraft, including aid to Communist China from 1949 through 1951. Notes on the current strength of the Red Chinese Air Force, aircraft types and insignia.

CHASSIN, LIONEL M. *Communist Conquest of China: A History of the Civil War, 1945–1949.* Cambridge, Mass.: Harvard University Press, 1965. Based on Red Chinese sources covering various campaigns and tactics, including the limited use of air power. Translated from French by Timothy Osato and Louis Gelas.

CHENG, CHU-YUAN. *Scientific and Engineering Manpower in Communist China, 1949–1963.* Washington, D.C.: Government Printing Office, 1965. A comprehensive analysis of engineering, training, and the application of engineering manpower in Communist China. Important notes on Soviet technical aid, the status of aerospace and aeronautical engineerinig in Red China, self-reliance, the production of military aircraft, and the foundation and faculties of aeronautical establishments, particularly the Peking Aeronautical Engineering College.

CHENNAULT, CLAIRE L. *Way of a Fighter*. New York: G. P. Putnam's Sons, 1949. Excellent survey of aviation in China in the 1930's and 1940's by the commander of the American Volunteer Group, that is, The Flying Tigers. Extensive discussion of the threat of a Communist take-over in China and of the early history and the potentialities of Communist Chinese use of air power.

"China Traders Thrive Again," *Business Week* (September 24, 1966). Discusses current foreign trade opportunities in Red China. Reviews Communist Chinese buying practices.

DOLLFUS, CHARLES, AND HENRI BOUCHE. *Histoire de L'Aéronautique (History of Aeronautics)*. Paris: L'Illustration, 1942. A pictorial history of aviation. Unique details on China between 1919 and 1939, including the war lord period.

DWIGGINS, DONALD. "Heros of the Hump," *Argosy* (February, 1966). Deals with CNAC flights over the Hump to India during World War II, postwar readjustment, and the mass defection of CNAC pilots and aircraft to the Chinese Communist side at the end of the Civil War.

ELKINS, H. R., AND THEON WRIGHT. *China Fights for Her Life*. New York: McGraw-Hill, 1938. Covers early conflict between Nationalist and Communist Chinese and the first months of the Sino-Japanese war. References to aviation.

ETHERTON, COLONEL P. T., AND H. HESSELL TILTMAN. *Manchuria—The Cockpit of Asia*. New York: Frederick A. Stokes, 1932. Detailed discussion of the Japanese invasion of Manchuria in 1931 and the threat of Chinese Communism. Occasional references to Japanese, Nationalist Chinese, and potential Red Chinese air power on the Asiatic mainland.

FUTRELL, ROBERT F., and LAWSON S. MOSELEY. *The United States Air Force in Korea: 1950–1953*. New York: Duell, Sloan & Pearce, 1961. Discusses the air war in Korea. Substantial coverage of Red Chinese participation, Russian aid, Chinese losses, and PLAAF strength. Based largely on Red Chinese sources and captured documents.

GENKO, UCHIDA. "Technology in China," *Scientific American* (November, 1966). Review of Red China's industrial base and growth. References to aircraft production.

GITTINGS, JOHN. *The Role of the Chinese Army*. London: Oxford University Press, 1967. Deals with PLA. Contains also important facts about the PLAAF, air force personalities, and aircraft strength from 1945 to the early 1960's. Mostly based on Communist Chinese sources.

GREEN, WILLIAM. *Fighters*. Vol. III (1961). New York: Hanover House. Third volume in a series on World War II aircraft. Includes Soviet fighter aircraft and reference to their export and foreign use.

———. *World Guide to Combat Planes.* Vol. I (1967). New York: Doubleday, 1967. Brief description of the PLAAF in 1966. Reference to current Soviet aircraft.

——— AND JOHN FRICKER. *Air Forces of the World.* London: MacDonald, 1958. History of all air forces of the world. Includes 1957 inventory of aircraft types for each nation. Large portions on Red China and Japan written by Richard M. Bueschel. Dr. Denys J. Voaden and Jacques Marmain wrote material on Russia.

——— and GERALD POLLINGER. *Aircraft of the World.* New York: Hanover House, 1956. A global inventory of aircraft in flying condition. Includes types supplied to Red China, primarily by the Soviet Union.

——— and D. PUNNETT. *World's Fighting Planes* (4th ed.). New York: Doubleday, 1964. A comprehensive survey of current combat aircraft. Includes Russian types sent to Red China and North Viet-Nam.

GRIFFITHS, SAMUEL B. II. "China's Capacity To Make War," *Foreign Affairs,* XLIII (January, 1965). Discusses Communist China's military capacity in 1964. Includes notes on Red Chinese air power. Based on radio intercepts and PLAAF documents.

HERMES, WALTER G. *Truce Tent and Fighting Front: United States Army in the Korean War.* Washington, D.C.: Government Printing Office, 1966. The second volume in a planned five-volume series on the U.S. Army in the Korean War. Reference to Soviet aircraft and training aid to the PLAAF, also to the air war over Korea, PLAAF strength, and the use of aircraft. Details on the growth of the PLAAF's strength in transport, and jet bombers.

HOBBS, LISA. *I Saw Red China.* New York: McGraw-Hill, 1966. Report on Red China by an American journalist. Observations on the state of civil aviation in Communist China in 1965.

HOOFTMAN, HUGO. *Russian Aircraft.* Fallbrook, Calif.: Aero Publishers, 1964. Excellent review of Russian aircraft of the 1950's and 1960's. Includes extensive references to the production and use of various Soviet types in Red China.

"Intelligence Report—Recruiting Engineers," *Parade* (April 30, 1967). Discusses Red Chinese recruitment of German aircraft engineers.

LEE, ASHER. *Soviet Air Force* (3d ed.). London: Gerald Duckworth, 1961. History of the Soviet Air Force. Frequent reference to Soviet aeronautical aid to Red China during and after the Korean War.

LEONARD, ROYAL. *I Flew for China.* New York: Doubleday, 1942. Memoirs of the pilot of Chang Hsueh-liang and Chiang Kai-shek. Reference to the frequent air contacts between Chang's Manchurian forces and the Communist forces at Sian and Yenan in the summer of 1936.

LIU, F. F. *A Military History of Modern China, 1924–1949*. Princeton, N.J.: Princeton University Press, 1956. A lucid military history of modern China. Contains one of the most extensive bibliographies on the Chinese military, including Nationalist Chinese, Communist Chinese, Russian, Japanese, English, German, and French sources.

NEMECEK, V. *Ceskoslovensko Letadla*. Prague: 1958. Review of aircraft produced in Czechoslovakia. Reference to aircraft exported to Red China.

NOWARRA, HEINZ J. *Die Sowjetischen Flugzeuge, 1941–1966 (Aircraft of the Soviet Union, 1941–1966)*. Munich: J. F. Lehmanns Verlag, 1967. Detailed coverage of Soviet military and civil aircraft from 1941 through 1966. Occasional references to aircraft exported to the People's Republic of China.

O'BALLANCE, EDGAR. *The Red Army of China: A Short History*. New York: Frederick A. Praeger, 1963. Contains notes on Red China's air power and its application. Main emphasis on period after World War II and the Chinese Civil War.

OMORI, MINORU. "Hanoi's Feuding Friends," *The New Republic* (October 15, 1966). Japanese report on current conditions in China. Reference to Red Chinese pride in producing "the latest model MIG-21."

ORLEANS, LEO A. "Research and Development in Communist China," *Science* (July 28, 1967). Review of research and development in Red China as related to defense needs. Details on the Academy of Military Sciences, aircraft, missiles, weapons development, and production requirements.

PORTISCH, HUGO. "China: Behind the Upheaval," *The Saturday Review* (December 10, 1966). Austrian journalist's report on Red China during the early period of the Cultural Revolution. Includes notes on the People's Revolution Military Museum and the American-built Lockheed U-2 display.

REA, GEORGE B. *The Case for Manchoukuo*. New York: D. Appleton-Century, 1935. Attempts to justify the Japanese occupation of Manchuria. Brief notes on aviation in Manchoukuo in the 1930's.

RHOADS, EDWARD J. M. *Chinese Red Army, 1927–1963: An Annotated Bibliography*. Cambridge, Mass.: Harvard University Press, 1964. Includes annotated list of the Red Chinese Air Force and Navy.

ROWE, DAVID N. *China Among the Powers*. New York: Harcourt, Brace, 1945. Discusses China's future after World War II, including China's potential imperialistic threat to Asia. Some references to Chinese air power.

SCHRAM, STUART R. (ed.). *Quotations from Chairman Mao-Tse-tung*. New York: Frederick A. Praeger, 1968. The famous "little red book." Self-view of the Chinese Communists. Occasional comments on Communist China's air power. Translated from the Chinese. Includes an introduction by Stuart R. Schram and a Foreword by A. Doak Barnett.

SNOW, EDGAR. *Red Star over China.* New York: Random House, 1938. A comprehensive review of the Chinese Communist movement, up to the fall of 1936, as seen by Mao Tse-tung and other veterans of the Long March. Occasional references to aviation and to the frequent air contacts between the Communists at Yenan and the Manchurian Tungpei Army in the spring and summer of 1936.

———. *The Pattern of Soviet Power.* New York: Random House, 1945. A book of great insight. Accurately predicted the extent of Soviet aid to Red China in the 1950's. References to aviation.

STEWART, JAMES T. *Airpower: The Decisive Force in Korea.* Princeton, N.J.: Van Nostrand, 1957. A detailed history of the use of American air power in Korea. Includes references to air war over North Korea and the participation by the PLAAF.

"The Strenuous Combat over the Wide and Lengthy Sky," *Chinese Air Force Magazine* (Taipei), 1959. A Nationalist Chinese paperback on the aerial combat over the Formosa Strait in 1958. In Chinese.

THOMAS, LOWELL, Jr. *The Silent War in Tibet.* New York: Doubleday, 1959. Detailed discussion of the Communist Chinese takeover of Tibet. Red Chinese sources. Includes references to PLAAF actions.

TREGASKIS, RICHARD. *Last Plane to Shanghai.* New York: Popular Library, 1962. An adventure novel about the Communist advance in China in the summer of 1948. Based on actual experience. Comments on the lack of Red Chinese air power during the Civil War.

UNISHEVSKY, VLADIMIR. *Red Pilot.* London: The Right Book Club, 1940. Memoirs of a defected Soviet pilot. References to Soviet aeronautical aid to foreign Communist movements. Interesting parallel to defections by Red Chinese pilots to Formosa.

WAGNER, RAY. *The North American Sabre.* New York: Doubleday, 1963. A complete history of the F-86 line of fighter aircraft. Extensive coverage of air fighting during the Korean War. Notes on PLAAF participation, aircraft, and losses.

WHITING, ALLEN S., AND GENERAL SHENG SHIH-TS'AI. *Sinkiang: Pawn or Pivot.* East Lansing, Mich.: Michigan State University Press, 1958. A history of the Sino-Soviet conflicts of interest along the northwest border of the province of Sinkiang. Reference to the use of aircraft.

YAKOVLEV, ALEXANDER. *Notes of an Aircraft Designer.* Translated by Albert Zdornykh. Moscow: Foreign Languages Publishing

House, 1960. Autobiography of Alexander Yakovlev. Description of his aircraft designs, many of which were exported to Red China.

MAGAZINES, NEWSPAPERS, AND PERIODICALS

Aerospace International. Washington, D.C.: United States Air Force Association. Bimonthly. Coverage of military and civil aviation, tactics, and air force strengths.

Air Britain Digest. Surrey, England: The International Association of Aviation Historians. Aviation history. References to civil and military aviation in Red China.

Air Classics. Bimonthly magazine on aviation history. Occasional references to Red Chinese aviation, such as the Korean War.

Aireview (Tokyo). Monthly. Aviation trade and general aviation matters. Japan's leading post-World War II aviation magazine. Extensive coverage of the war in Vietnam, Red China's air power, and Red Chinese civil aviation. Prepared by Japanese reporters in Red China.

Air Force and Space Digest (formerly *Air Force*). Washington, D.C.: Air Force Association. Monthly aerospace magazine. News and features on military aviation. Frequent reference to Red Chinese military aviation.

Air Pictorial (London). Monthly. History of aviation and general aviation matters. Frequent news of Red Chinese civil and military aviation.

Air Progress. Monthly. History of aviation and general aviation matters. Frequent references to Red Chinese civil and military aviation.

American Aviation. Monthly aviation trade magazine. Scattered references to Red Chinese civil aviation.

American Aviation Historical Society Journal. Whittier, Calif.: American Aviation Historical Society. Quarterly aviation history. Emphasis on the development and use of American aicraft in the world. Very detailed. Based primarily on original sources. Frequent reference to aviation and the use of American aircraft in China.

Asahigraph (The Rising Sun Graphic) (Tokyo). Vol. XXVI (No. 22), May 27, 1936–Vol. XL (No. 2), January 13, 1943. Weekly news magazine. Extensive coverage of Chinese aviation during the Sino-Japanese war and World War II.

Aviation Week and Space Technology (formerly *Aviation Week*). Frequent and original coverage of Red Chinese military and civil aviation and of advanced missile and rocket development.

China Aviation. Chungking: Chinese Association for Promotion of Aviation, March 31, 1939. Quarterly on progress in Chinese military and civil aviation. Reports on war lords, the Fukien Rebellion, and the Sino-Japanese War. Infrequently published. In Chinese and English.

The China Clipper (New York). Monthly journal of the China Stamp Society. Interesting insights into Chinese civil and airmail aviation since the Sino-Japanese War.

China Pictorial (Peking). Monthly propaganda magazine of general interest for foreign distribution. Frequent comments on Red Chinese civil aviation. Occasional references to the PLAAF.

The China Quarterly (London). Quarterly review of developments in the People's Republic of China. Articles on military developments and nuclear weapons. References to the PLA and Red Chinese military and civil aviation.

The China Weekly Review (Shanghai). Vols. XXV–CXIV. Weekly. Issues cited cover the period 1923 until immediately after the Communist take-over of Shanghai, in June, 1949. Extensive coverage of aeronautical developments in China. Good coverage of the use of Communist and Nationalist air power during the Civil War.

Chinese Air Force Magazine (formerly *China's Air Force*) (Taipei, formerly Chungking). Monthly. Sporadically published during the Sino-Japanese War. In the 1950's and 1960's extensive coverage of the aerial encounters between the PLAAF and the Nationalist Chinese Air Force.

Chinese Air Force Pictorial (Taipei). Yearbook. Often describes and illustrates defected Red Chinese aircraft and pilots.

Flying (formerly *Popular Aviation*). Monthly aviation magazine. Occasional references to China during World War II.

Flying Review International (formerly *Royal Air Force Flying Review*) (London). Monthly. Deals with current aeronautical events, historical aircraft, aviation news, aeronautical history, and trade news. Frequent and outstanding coverage of Red Chinese aviation.

Free China Review (Taipei). Monthly propaganda magazine for foreign distribution. Includes data on the PLAAF, its encounters with the Nationalist Chinese, its strength, and equipment.

Free China Weekly (Taipei). Propaganda magazine. References to PLAAF, its contacts with the Nationalist Chinese, and Red Chinese defections.

Jiefangjun Huabao or *Chieh-Fang-Chun Hua-Pao (Liberation Army Pictorial)* (Peking). Red Chinese military magazine of general interest. Frequent articles about the PLAAF.

Jiefangjun Pao or *Chieh-Fang-Chun Pao (Liberation Army Journal)* (Peking). Periodical for the People's Liberation Army. Frequent stories, photographs and historical data on PLAAF aircraft and events. Data require confirmation.

Koku Asahi (The Aerial Rising Sun) (Tokyo). Vols. I–VI (1940–45). Monthly aviation magazine. Coverage of the Sino-Japanese and Pacific Wars. Extensive coverage of aviation in occupied China, Manchoukuo, and the air war in China.

Koku Fan (Aero Fan) (Tokyo) Monthly. Covers general and model aviation. Frequent pictorial coverage of Communist Chinese aviation.

Koku Jidai (The Aerial Age) (Tokyo). Vols. I–XIII (1930–42). Monthly. General and trade aviation magazine. Extensive coverage of aviation in China.

Newsweek. Frequent references to Red Chinese military and civil aviation.

The New York Times. Frequent references to Red Chinese military and civil aviation.

Sekai no Kokuki (The World's Aircraft) (Tokyo). Nos. 1–80 (1951–57). Monthly. History of aviation and general aviation matters. References to the growth of military and civil aviation in the Far East. Extensive coverage of Red Chinese aviation, particularly during the Korean War.

Shashin Shuho (Photo Weekly) (Tokyo). (1938–45). Weekly military propaganda magazine of the Japanese Government. References to military and political events of the Sino-Japanese and the Pacific Wars. Extensive photographic coverage of the air war in China.

Skrzydlata Polska (Polish Aviation) (Warsaw). (1957–59). Leading magazine on general and military aviation. Comprehensive reports on new aircraft types produced in the People's Republic of China in the late 1950's.

Sora (Sky) (Tokyo). Vols. I–IX (1934–42). Monthly. Pictorial magazine on general aviation. Extensive coverage of the air war in China.

Sovetskaya Aviatsiya (Soviet Aviation) (Moscow). (1958–59). Popular Russian aviation magazine. Reports on the growth of the PLAAF after the Korean War from the Soviet point of view. Major credit is given to Russian aid and equipment.

Taiwan Pictorial (Taipei). Monthly magazine of general interest. Detailed coverage of PLAAF defections to South Korea and Formosa.

Time. Frequent coverage of Red Chinese military and civil aviation as part of reports on Communist China.

Umi-to-Sora (Sea and Air) (Tokyo). Vols. I–XIV (1932–45). Military pictorial magazine dealing with ships, tanks, and aircraft. Includes aeronautical events in China in the 1930's and 1940's.

U.S. News and World Report. Frequent articles on Red Chinese politics, economics, and military affairs. Periodic coverage of Red Chinese nuclear weapons, air power and military potential.

INDEX

A-3D Skywarrior bombers (Douglas), 63
ABC gliders (SZD IS-3), 37, 187–88
ABC-A gliders (SZD IS-3), 37, 38, 188
ABC-A gliders (Tchan Tia-kuo), 38, 194–95
Academy of Military Sciences, 70, 71, 88, 90
Aero L-29 Delfin trainers, 109
Aero Super-Aero 45 transports, 45, 53, 108–9, 126
Aeronautical Enterprises of the Czechoslovak Socialist Republic (Aero), 108
AFA P-16 Mark III fighters, 69
AI-20 engines (Ivchenko), 48–49
Air Force Academy, PLAAF (Sian), 19, 27, 32
Air France, 51
Albatross aircraft (Grumman HU-16), 55
Alouette III helicopters (Sud-Aviation SE-3160), 69, 186–87
Amoy Naval Establishment Aircraft Factory, 109
Amoy Pan trainers, 9, 109–10
Amphibian aircraft (Loening-Keystone), 9, 141
AN-2 (Fong Shou No. 2) utility aircraft (Shenyang), 41, 45, 47, 53, 82, 168–69, 177–78
AN-2 transports (Antonov), 34, 41, 73, 110–11
AN-2S transports (Antonov), 41
AN-14 transports (Antonov), 44, 167
AN-24 transports (Antonov), 124
Antonov, Oleg K., 110–11, 166
Antonov AN-2 transports, 34, 41, 73, 110–11
Antonov AN-2S transports, 41
Antonov AN-14 transports, 44, 167
Antonov AN-24 transports, 124
A6A attack planes (Grumman), 66
Automation and Remote Control, Institute of, 70
Aviakhim R-1.M-5 bombers, 5, 6, 7, 112–13
Aviakhim U-1 Avrushka trainers, 5, 112
Avian IV trainers (Avro), 8, 9, 113
Aviation Bureau, Nationalist Chinese, 5
Avro Avian IV trainers, 8, 9, 113
Avro 621 Tutor trainers, 114
Avrushka trainers (Aviakhim U-1), 5, 112

B-24J Liberator bombers (Consolidated), 15, 117–18
B-25H Mitchell bombers (North American), 18, 19, 162

B-25J Mitchell bombers (North American), 162
B-47 Stratojet bombers (Boeing), 59
B-57 bombers (Martin), 71, 85
Baade, Brunolf, 204–5, 206
BAC-111 transports, 50
"Badger" (code name), 204
"Bank" (code name), 162
"Bark" (code name), 128
"Bat" (code name), 200
BE-6 flying boats (Beriev), 34, 114
"Beagle" (code name), 134
"Beast" (code name), 130
Beaver transports (De Havilland-Canada DHC-2), 71, 120–21
Bei-Jing No. 1 transports, 44, 53, 165–66
Beriev, Georgi M., 115
Beriev BE-6 flying boats, 34, 114
BOAC Bristol Britannia 102 airliners, 50
Bocian sailplanes (SZD-9), 37, 190–91
Boeing B-47 Stratojet bombers, 59
Boeing PT-17 Kaydet trainers, 18, 115–16
Boeing 720B transports, 49
Borodin, Mikhail, 5, 6, 7
"Bosun" (code name), 203
BQM-34A Firebee aircraft (Ryan), 87
Breguet 14/400 bombers, 116–17
Bristol Britannia 102 airliners (BOAC), 50
"Buck" (code name), 170
"Bull" (code name), 201
"Butcher" (code name), 134

C-46 Commando transports (Curtiss), 18, 26, 56, 119
C-47 Dakota transports (Douglas), 13, 18, 27, 48, 56, 122–23
"Cab" (code name), 141
Canton Military Region, 86
Capital No. 1 transports, 45
"Cart" (code name), 202
Central Air Transport Corporation (CATC), 45
Central Gliding School, Anyang, 37
Chang Chi-wei, 25
Chang Ting-fa, 95
Chao Teh-an, 54
Chen Wai-sheng, 31, 58
Cheng Chun, 80
Chia Jen-chu, 178
Chiang Kai-shek, 5, 6, 8, 9, 11, 12, 13, 79, 207
CHICOMAF, 3
Chien Hsueh-shen, 40, 41, 42, 70, 71, 165
China Aeronautical Engineering Society, 71
China National Aviation Corporation (CNAC), 45
China-Salamandra gliders (Tchan Tia-kuo), 38, 196–97
Chinese Academy of Sciences, 70
Chinese Government Purchasing Corporation, 52
Chinese Ministry of Aviation, 52
Chinese People's Liberation Army Naval Air Force (PLANAF), 21, 34, 64, 73, 74, 94, 97, 99, 100, 102, 103
Chinko No. 1 general-purpose aircraft (Shenyang), 42, 179–80
Chipmunk trainers (De Havilland-Canada DHC-1), 120
Cho Yung, 80
Chou En-lai, 11, 13, 49, 96, 133
Chu Teh, 8, 10, 11
Chu Wu-hua, 127
Chung-Jun transports, 44
Civil Aviation Administration of China (CAAC), 46, 47. 48, 50, 52, 95, 96
"Clod" (code name), 168
"Coach" (code name), 131
"Colt" (code name), 111, 178
Commando transports (Curtiss C-46), 18, 26, 56, 119
Consolidated B-24J Liberator bombers, 15, 117–18
Convair PBY-5A aircraft, 55
"Cork" (code name), 214
"Crate" (code name), 132
"Creek-A" (code name), 180
"Creek-B" (code name), 212
"Creek-C" (code name), 212
"Creek-D" (code name), 212
"Crow" (code name), 212
Cultural Revolution, 93, 96, 98

Curtiss C-46 Commando transports, 18, 26, 56, 119
Curtiss Export Hawk II fighters, 9, 118–19

Dakota transports (Douglas C-47), 13, 18, 27, 48, 56, 122–23
Dalai Lama, 141
Dassault Mirage fighters, 69
DC-130A Hercules transports (Lockheed), 87
De Havilland–Canada DHC-1 Chipmunk trainers, 120
De Havilland–Canada DHC-2 Beaver transports, 71, 120–21
De Havilland DH-9A bombers, 112
De Havilland DH-98 Mosquito bombers, 18, 119–20
Delfin trainers (Aero L-29), 109
Denikin, General, 112
Destroyer aircraft (Douglas RB-66), 57, 61
DH-9A bombers (De Havilland), 112
DH-98 Mosquito bombers (De Havilland), 18, 119–20
DH-121 Trident transports (Hawker-Siddeley), 50
DHC-1 Chipmunk trainers (De Havilland–Canada), 120
DHC-2 Beaver transports (De Havilland–Canada), 71, 120–21
Doolittle, James, 118
Douglas A-3D Skywarrior fighters, 63
Douglas C-47 Dakota transports, 13, 18, 27, 48, 56, 122–23
Douglas O-2MC4 Marx general-purpose aircraft, 10, 121–22
Douglas RB-66 Destroyer aircraft, 57, 61
Draken fighters (SAAB J-35B), 69

E13A1 Type O seaplanes (Watanabe), 208–9
843 Viscount transports (Vickers), 50, 52, 124, 206–7
Experimental Glider Establishment (Poland), 191, 194
Export Hawk II fighters (Curtiss), 9, 118–19
F-4B Phantom fighters (McDonnell), 62
F-4C Phantom fighters (McDonnell), 61
F.13 transports (Junkers-Fili), 5, 135–36
F-86 Sabre fighters (North American), 21, 23, 24, 85, 162–63
F-86F Sabre fighters (North American), 54, 71, 162–64
F-104 Starfighters (Lockheed), 63, 71, 85, 94
"Faceplate" (code name), 149
"Fagot" (code name), 144, 145
"Falcon" (code name), 144
"Fang" (code name), 139
"Fargo" (code name), 143
"Farmer" (code name), 149, 180
"Feather" (code name), 215
Feilung Hiryu No. 1 seaplanes, 43, 123
Feilung Machine Building Works, 43, 123
Fi. 156 Storch aircraft (Fieseler), 111
Fieseler Fi. 156 Storch aircraft, 111
Fifth Extermination Campaign, 9
53A trainers (SZD Salamandra), 37, 38, 188–89
Firebee aircraft (Ryan BQM-34A), 87
"Fishbed" (code name), 149
"Fishbed-B" (code name), 150, 152, 182
"Fishbed-C" (code name), 151, 152
"Fishbed-D" (code name), 153
"Fishbed-F" (code name), 151, 153
"Fishbed-G" (code name), 151
"Flying Dragons," 43, 123
Fokker-Fairchild Friendship transports, 124
Fong Shou No. 2 (*see* Shenyang AN-2 utility aircraft)
45 transports (Super-Aero), 45, 53, 108–9, 126
14/400 bombers (Breguet), 116–17
"Frank" (code name), 160, 210
"Fresco-A" (code name), 147
"Fresco-B" (code name), 147
"Fresco-C" (code name), 147, 175
"Fresco-D" (code name), 147

Freytag, Fritz, 204–6
Friendship transports (Fokker-Fairchild), 124
"Fritz" (code name), 138
Fukien Rebellion, 9

Galen, Vassily, 6, 7
Gliders, 37–38
Gliding Institute (Poland), 187
Great Leap Forward, 47, 48
Grumman A6A attack planes, 66
Grumman HU-16 Albatross aircraft, 55
Gurevich, Mikhail I., 143

Hai Lun-Kiang No. 1 agricultural planes, 43, 125–26
Handley Page Company, 124
Handley Page HPR7 Herald transports, 53, 124–25
Harbin Aeronautical Engineering College, 125, 126
Harbin Aeronautical Engineering Works, 43, 45, 126–28
Harbin Polytechnic University, 71, 125
"Hare" (code name), 154
Hawk II fighters (Curtiss Export), 9, 118–19
Hawker-Siddeley DH-121 Trident transports, 50
Hayabusa fighters (Nakajima Ki.43 Type 1), 14, 158–59
Hayate fighters (Nakajima Ki.84 Type 4), 14, 159–60
Herald transports (Handley Page HPR7), 53, 124–25
Hercules transports (Lockheed DC-130A), 87
"Hickory" (code name), 192, 193
Hien fighters (Kawasaki Ki.61 Type 3), 14, 136–37
Hiryu No. 1 seaplanes (Feilung), 43, 123
"Hound" (code name), 155, 179
HPR7 Herald transports (Handley Page), 53, 124–25
Hsu Ting-tse, 79
HU-16 Albatross aircraft (Grumman), 55
Hurley, Patrick J., 12–13
Hwang Chi-hsiang, 6
Hydrogen bomb, Chinese explosion of, 99

"Ida" (code name), 194
IL-2m3 bombers (Ilyushin), 19, 127–28
IL-10 bombers (Ilyushin), 19, 21, 26, 128–29
IL-12 transports (Ilyushin), 19, 27, 34, 73, 130–31
IL-14 transports (Ilyushin), 73, 75, 84, 131
IL-14M transports (Ilyushin), 46, 52, 131–32
IL-14P transports (VEB Flugzeugwerke Dresden), 47, 52, 53, 204–6
IL-18 transports (Ilyushin), 48–49, 51, 73, 132–33
IL-28 bombers (Ilyushin), 22, 32, 33, 36, 73, 74, 79, 81, 83, 85, 100, 133–34
IL-28U trainers (Ilyushin), 33, 73, 81, 134–35
Ilyushin, Sergei V., 130
Ilyushin IL-2m3 bombers, 19, 127–28
Ilyushin IL-10 bombers, 19, 21, 26, 128–29
Ilyushin IL-12 transports, 19, 27, 34, 73, 130–31
Ilyushin IL-14 transports, 73, 75, 84, 131
Ilyushin IL-14M transports, 46, 52, 131–32
Ilyushin IL-18 transports, 48–49, 51, 73, 132–33
Ilyushin IL-28 bombers, 22, 32, 33, 36, 73, 74, 79, 81, 83, 85, 100, 133–34
Ilyushin IL-28U trainers, 33, 73, 81, 134–35
Institute of Mechanics and Electronics, 70
Institute of Mechanics of the Chinese Academy of Sciences, 40, 70
Institute of Upper Atmospheric Physics, 70
IS-3 ABC gliders (SZD), 37, 187–88
IS-3 ABC-A gliders (SZD), 37, 38, 188
IS-4 Jastraab sailplanes (SZD), 37, 189–90
Ivchenko AI-20 engines, 48–49

J-35B Draken fighters (SAAB), 69
"Jake" (code name), 209
Jaskolka sailplanes (SZD-8), 37, 190
Jaskolka sailplanes (Tchan Tia-kuo SZD-8), 38, 195
Jastraab sailplanes (SZD IS-4), 37, 189–90
Jie-Fang No. 1 (Liberation No. 1) sailplanes (Tchan Tia-kuo), 38, 197–98
Jouett, John J., 207
Junkers-Fili F.13 transports, 5, 135–36

Kao Chang-chi, 58
Kao Yu-tsung, 77, 78
Kawasaki Ki.48 Type 99 bombers, 14, 136
Kawasaki Ki.61 Type 3 Hien fighters, 14, 136–37
Kaydet trainers (Boeing PT-17), 18, 115–16
Khrushchev, Nikita, 42, 67, 132–33
Ki.34 Type 97 transports (Nakajima), 14, 157–58
Ki.43 Type 1 Hayabusa fighters (Nakajima), 14, 158–59
Ki.44 Type 2 Shoki fighters (Nakajima), 14, 159
Ki.48 Type 99 bombers (Kawasaki), 14, 136
Ki.51 Type 99 bombers (Mitsubishi), 14, 155–56
Ki.54a Type 1 trainers (Tachikawa), 14, 20, 192
Ki.54c Type 1 transports (Tachikawa), 14, 192–93
Ki.55 Type 99 trainers (Tachikawa), 14, 15, 20, 193–94
Ki.57 Type 100 transports (Mitsubishi), 14, 156–57
Ki.61 Type 3 Hien fighters (Kawasaki), 14, 136–37
Ki.79 Type 2 trainers (Manshu), 14, 142
Ki.84 Type 4 Hayate fighters (Nakajima), 14, 159–60
KMT (*see* Kuomintang)
Kolesnikov, D. N., 213
Kulishenko, Grigori A., 11
Kunchangtang (KCT), 4
Kuo Yung-hua, 71
Kuo Yung-huai, 165
Kuomintang (KMT), 5, 6, 7, 8

L-29 Delfin trainers (Aero), 109
LA-9 fighters (Lavochkin), 19, 26, 33, 35, 73, 137–38
LA-11 fighters (Lavochkin), 19, 20, 26, 33, 73, 138–39
LA-15 fighters (Lavochkin), 24, 139–40
Lavochkin, Semyon A., 139–40
Lavochkin LA-9 fighters, 19, 26, 33, 35, 73, 137–38
Lavochkin LA-11 fighters, 19, 20, 26, 33, 73, 138–39
Lavochkin LA-15 fighters, 24, 139–40
Lenin general-purpose aircraft (Vought V-65-C1), 10, 207–8
Li Han, 25
Li Hsien-pin, 79, 80
Li Tsai-wang, 79
LI-2 transports (Lisunov), 19, 27, 34, 46, 48, 73, 82, 140–41
Liang Kuo-hsin, 80
Liberation No. 1 (*see* Jie-Fang No. 1 sailplanes)
Liberator bombers (Consolidated B-24J), 15, 117–18
Lien Pao-sheng, 79, 80
"Lily" (code name), 136
Lin Piao, 95
Lisunov LI-2 transports, 19, 27, 34, 46, 48, 73, 82, 140–41
Liu Cheng-sze, 78
Liu Pu-tung, 164
Liu Shan-pan, 118
Liu Shao-chi, 49, 95, 96
Liu Ya-lou, 18, 82, 83, 89
Lockheed DC-130A Hercules transports, 87
Lockheed F-104 Starfighters, 63, 71, 85, 94
Lockheed RT-33 aircraft, 57
Lockheed SR-71 reconnaissance aircraft, 97
Lockheed U-2 reconnaissance aircraft, 57, 58, 96
Loening, Grover, 141

Loening-Keystone Amphibian aircraft, 9, 141
"Long March," 10
Lum Wai-sing, 6

McDonnell F-4B Phantom fighters, 62
McDonnell F-4C Phantom fighters, 61
McDonnell RF-101 reconnaissance aircraft, 57, 59
"Madge" (code name), 115
"Magnet" (code name), 215
Manshu Airplane Manufacturing Company, 39, 125, 142, 175
Manshu Ki.79 Type 2 trainers, 14, 142
Mao An-ying, 24
Mao Tse-tung, 4, 8, 9, 10, 13, 16, 19, 24, 25, 29, 47, 53, 67, 92, 93, 95, 104, 149
Mare" (code name), 212
Mark III fighters (AFA P-16), 69
Martin B-57 bombers, 71, 85
Martin RB-57D aircraft, 57
Marx general-purpose aircraft (Douglas O-2MC4), 10, 121–22
"Mascot" (code name), 135
"Max" (code name), 175, 216
MI-1 helicopters (Mil), 20, 154
MI-4 helicopters (Mil), 34, 38, 42, 155
MI-4 helicopters (Shenyang), 42, 75
"Midget" (code name), 146
MiG-9 fighters (Mikoyan), 19, 142–43
MiG-15 fighters (Mikoyan), 20, 21, 23, 24, 26, 33, 57, 73, 143–44
MiG-15bis fighters (Mikoyan), 24, 33, 34, 53, 78, 81, 82, 83, 85, 91, 100, 144–45
MiG-15UTI trainers (Mikoyan), 33, 73, 145–46
MiG-15UTI trainers (Shenyang), 40, 85, 176–77
MiG-17 fighters (Mikoyan), 31, 33, 34, 36, 54, 56, 57, 61, 63, 65, 82, 83, 86, 91, 100, 146–47
MiG-17 fighters (Shenyang), 40, 41, 53, 59, 67, 73, 81, 85, 87, 175–76
MiG-19 fighters (Mikoyan), 57, 58, 66–67, 87, 91, 100, 147–49
MiG-19 fighters (Shenyang), 67, 68, 84, 85, 87, 88, 94, 180–81
MiG-21 fighters (Mikoyan), 83, 87–88, 149–50
MiG-21 fighters (Shenyang), 89, 90, 92, 93, 100, 181–83
MiG-21f fighters (Mikoyan), 88, 90, 91, 150–52
MiG-21Pf fighters (Mikoyan), 90, 91, 152–53
Mikoyan, Artem I., 143
Mikoyan MiG-9 fighters, 19, 142–43
Mikoyan MiG-15 fighters, 20, 21, 23, 24, 26, 33, 57, 73, 143–44
Mikoyan MiG-15bis fighters, 24, 33, 34, 53, 78, 81, 82, 83, 85, 91, 100, 144–45
Mikoyan MiG-15UTI trainers, 33, 73, 145–46
Mikoyan MiG-17 fighters, 31, 33, 34, 36, 54, 56, 57, 61, 63, 65, 82, 83, 86, 91, 100, 146–47
Mikoyan MiG-19 fighters, 57, 58, 66–67, 87, 91, 100, 147–49
Mikoyan MiG-21 fighters, 83, 87–88, 149–50
Mikoyan MiG-21f fighters, 88, 90, 91, 150–52
Mikoyan MiG-21Pf fighters, 90, 91, 152–53
Mil, Mikhail L., 154
Mil MI-1 helicopters, 20, 154
Mil MI-4 helicopters, 34, 38, 42, 75, 155
Military Air Command (MAC), 101
Military Science Academy, 97, 98, 183
Ministry of Machine Building, 38, 39, 72, 89, 97, 183–84
Mirage fighters (Dassault), 69
"Mist" (code name), 213
Mitchell bombers (North American B-25H and B-25J), 18, 19, 162
Mitsubishi Ki.51 Type 99 bombers, 14, 155–56
Mitsubishi Ki.57 Type 100 transports, 14, 156–57
"Moose" (code name), 211

Mosquito fighters (De Havilland DH-98), 18, 119–20
Mucha-100 sailplanes (SZD-12), 37, 191–92
Mucha-100 sailplanes (Tchan Tia-kuo SZD-12), 38, 195–96
"Mule" (code name), 173
Mustang fighters (North American P-51D), 18, 19, 160–61

Nakajima Ki.34 Type 97 transports, 14, 157–58
Nakajima Ki.43 Type 1 Hayabusa fighters, 14, 158–59
Nakajima Ki.44 Type 2 Shoki fighters, 14, 159
Nakajima Ki.84 Type 4 Hayate fighters, 14, 159–60
Nanking Aeronautical Engineering College, 71
National Aircraft Factory (Shenyang), 39, 40, 41, 67, 70, 85, 93, 97, 184
National Defense Council, 72
Nationalist Chinese Air Force, 15, 16, 17, 35, 54, 94, 102
Naval Air Establishment (NAE), 44, 123
Neptune reconnaissance aircraft (P2V-7), 57
Nicolle, S. J., 124
Niespala, J., 37, 38, 197–98
North American B-25H Mitchell bombers, 18, 19, 162
North American B-25J Mitchell bombers, 162
North American F-86 Sabre fighters, 21, 23, 24, 85, 162–63
North American F-86F Sabre fighters, 54, 71, 162–64
North American P-51D Mustang fighters, 18, 19, 160–61
North Korean Air Force, 81
Northwestern Technological University, 43, 71, 164–65
Nur Khan, M., 52

O-2MC4 Marx general-purpose aircraft (Douglas), 10, 121–22
102 airliners (BOAC Bristol Britannia), 50
"Oscar" (code name), 158

P-16 Mark III fighters (AFA), 69
P-47D Thunderbolt fighters (Republic), 18, 173–74
P-51D Mustang fighters (North American), 18, 19, 160–61
Pacific Air Forces (PACAF), 101
Pakistan International Airlines (PIA), 49
Pakistani Air Force, 84, 181
Pan, D. S., 109
Pan trainers (Amoy), 9, 109–10
Pao Shou-ken, 54
PBY-5A aircraft (Convair), 55
PE-2 bombers (Petlyakov), 19, 169–70
Peking Aeronautical Engineering College, 44, 45, 70, 98, 165–69
People's Daily (Peking), 48
People's Liberation Army (PLA), 15–18
People's Liberation Army Air Force (PLAAF), 3, 4, 18, 19, 20, 22, 23, 25, 26, 27, 30, 31, 32, 33, 34, 35, 36, 37, 38, 40, 41, 42, 49, 53, 55, 57, 58, 64, 65, 66, 69, 70, 72, 73, 75, 76, 77, 83, 87, 90, 92, 93, 94, 95, 97, 99, 100, 101, 102, 103, 104
Petlyakov, Vladimir M., 170
Petlyakov PE-2 bombers, 19, 169–70
Phantom fighters (McDonnell F-4B and F-4C), 61, 62
Philip, Duke of Edinburgh, 124
PLAAF (*see* People's Liberation Army Air Force)
PLANAF (*see* Chinese People's Liberation Army Naval Air Force)
PO-2 utility aircraft (Polikarpov), 20, 73, 111, 172–73
Polikarpov, Nikolai N., 172
Polikarpov PO-2 utility aircraft, 20, 73, 111, 172–73
Polikarpov R-5 general-purpose aircraft, 11, 170–71
PT-17 Kaydet trainers (Boeing), 18, 115–16
P2V-7 Neptune reconnaissance aircraft, 57

R-1.M-5 bombers (Aviakhim), 5, 6, 7, 112–13
R-5 general-purpose aircraft (Polikarpov), 11, 170–71
R-6 helicopters (Sikorsky), 185–86
RB-57D reconnaissance aircraft (Martin), 57
RB-66 Destroyer reconnaissance aircraft (Douglas), 57, 61
Red Banner No. 1 transports, 45, 168–69
Red Guards, 93
Republic P-47D Thunderbolt fighters, 18, 173–74
Republic RF-84 reconnaissance aircraft, 57
Republic RF-84F Thunderflash reconnaissance fighters, 31
RF-84 reconnaissance aircraft (Republic), 57
RF-84F Thunderflash reconnaissance fighters (Republic), 31
RF-101 reconnaissance aircraft (McDonnell), 57, 59
Roderick, John, 16
Rogachev, Comrade, 5
Royal Khmer Aviation, 85
RT-33 reconnaissance aircraft (Lockheed), 57
Ryan BQM-34A Firebee aircraft, 87

SAAB J-35B Draken fighters, 69
Sabre fighters (North American F-86 and F-86F), 21, 23, 24, 54, 71, 85, 162–64
Salamandra 53A trainers (SZD), 37, 38, 188–89
SB-2 bombers (Tupolev), 11, 198–99
SE-210 transports (SUD-Caravelle), 51
SE-3160 Alouette III helicopters (Sud-Aviation), 69, 186–87
720B transports (Boeing), 49
Seventh Fleet, U.S., 28, 29, 53, 74, 101
Sha-Tu No. 1 transports, 45, 166–68
Shao Hsi-yen, 77, 78
Shen Yuan, 71
Shenyang Aeronautical Engineering College, 179
Shenyang AN-2 (Fong Shou No. 2) utility aircraft, 41, 45, 47, 53, 82, 168–69, 177–78
Shenyang aviation school, 43
Shenyang Chinko No. 1 general-purpose aircraft, 42, 179–80
Shenyang medium bombers, 184–85
Shenyang MI-4 helicopters, 42, 75
Shenyang MiG-15UTI trainers, 40, 85, 176–77
Shenyang MiG-17 fighters, 40, 41, 53, 59, 67, 73, 81, 85, 87, 175–76
Shenyang MiG-19 fighters, 67, 68, 84, 85, 87, 88, 94, 180–81
Shenyang MiG-21 fighters, 89, 90, 92, 93, 100, 181–83
Shenyang multi-purpose fighters, 183–84
Shenyang School of Aeronautical Industry, 179
Shenyang Whirlwind-25 helicopters, 42, 178–79
Shenyang YAK-18 trainers, 174–75
Shoki fighters (Nakajima Ki.44 Type 2), 14, 159
Shvetsov, A. D., 115
Sihanouk, Norodom, 86
Sikorsky, Igor I., 154
Sikorsky R-6 helicopters, 185–86
621 Tutor trainers (Avro), 114
Skywarrior bombers (Douglas A-3D), 63
"Sonia" (code name), 156
Special Flight Group, 41
SR-71 reconnaissance aircraft (Lockheed), 97
Stalin, Joseph, 5, 20, 42
Starfighters (Lockheed F-104), 63, 71, 85, 94
Storch aircraft (Fieseler Fi. 156), 111
Strategic Air Command (SAC), 101
Stratojet bombers (Boeing B-47), 59
Sud-Aviation SE-3160 Alouette III helicopters, 69, 186–87
SUD-Caravelle SE-210 transports, 51
Sun Yat-sen, 5, 7, 79
Sungari No. 1 transports, 45, 126–27
Super-Aero 45 transports, 45, 53, 108–9, 126

SZD-8 Jaskolka sailplanes (SZD), 37, 190
SZD-8 Jaskolka sailplanes (Tchan Tia-kuo), 38, 195
SZD Gliding Institute, 37
SZD IS-3 ABC gliders, 37, 187–88
SZD IS 3 ABC-A gliders, 37, 38, 188
SZD IS-4 Jastraab sailplanes, 37, 189–90
SZD-9 Bocian sailplanes, 37, 190–91
SZD Salamandra 53A trainers, 37, 38, 188–89
SZD-12 Mucha-100 sailplanes (SZD), 37, 191–92
SZD-12 Mucha-100 sailplanes (Tchan Tia-kuo), 38, 195–96
Szybowcowy Zaklad Doswiad Czalny (SZD), 37

Tachikawa Ki.54a Type 1 trainers, 14, 20, 192
Tachikawa Ki.54c Type 1 transports, 14, 192–93
Tachikawa Ki.55 Type 99 trainers, 14, 15, 20, 193–94
Tchan Tia-kuo ABC-A gliders, 38, 194–95
Tchan Tia-kuo China-Salamandra gliders, 38, 196–97
Tchan Tia-kuo design office, 38
Tchan Tia-kuo Glider Manufacturing Center, 194
Tchan Tia-kuo gliders, 197
Tchan Tia-kuo Jie-Fang No. 1 (Liberation No. 1) sailplanes, 38, 197–98
Tchan Tia-kuo SZD-8 Jaskolka sailplanes, 38, 195
Tchan Tia-kuo SZD-12 Mucha-100 sailplanes, 38, 195–96
"Thora" (code name), 158
Thunderbolt fighters (Republic P-47D), 18, 173–74
Thunderflash fighters (Republic RF-84F), 31
"Tojo" (code name), 159
"Tony" (code name), 137
"Topsy" (code name), 157
Trident transports (Hawker-Siddeley DH-121), 50
Tsai Ting-kai, 8
TU-2 bombers (Tupolev), 19, 21, 26, 32, 33, 34, 37, 73, 199–200
TU-4 bombers (Tupolev), 34, 60, 73, 74, 75, 100, 114, 200–201
TU-14 bombers (Tupolev), 35, 202–3
TU-16 bombers (Tupolev), 92, 203–4
TU-70 transports (Tupolev), 73, 201–2
Tung Hsiao-hai, 59, 63
Tupolev, Andrei N., 170, 199, 200
Tupolev SB-2 bombers, 11, 198–99
Tupolev TU-2 bombers, 19, 21, 26, 32, 33, 34, 37, 73, 199–200
Tupolev TU-4 bombers, 34, 60, 73, 74, 75, 100, 114, 200–201
Tupolev TU-14 bombers, 35, 202–3
Tupolev TU-16 bombers, 92, 203–4
Tupolev TU-70 transports, 73, 201–2
Tutor trainers (Avro 621), 114
Twenty-fifth Air Division of PLAAF, 75

U-1 Avrushka trainers (Aviakhim), 5, 112
U-2 reconnaissance aircraft (Lockheed), 57, 58, 96
USAF Tactical Air Force, 101
U.S. Navy P2V-7 Neptune reconnaissance aircraft, 57
U.S. Seventh Fleet, 28, 29, 53, 74, 101

V-65-C1 Lenin general-purpose aircraft (Vought), 10, 207–8
VC-10 transports (Vickers), 50
VEB Flugzeugwerke Dresden IL-14P transports, 47, 52, 53, 204–6
Vickers 843 Viscount transports, 50, 52, 124, 206–7
Vickers VC-10 transports, 50
Viscount transports (Vickers 843), 50, 52, 124, 206–7
Voroshilov, Colonel, 19
Vought V-65-C1 Lenin general-purpose aircraft, 10, 207–8

Wang Cheng, 72
Wang Ching-wei, 5

Wang Hai, 25
Wang Pin-cheng, 72, 83, 89
Watanabe E13A1 Type O seaplanes, 208–9
Wen Lin-tschen, 109–10
Whirlwind-25 helicopters (Shenyang), 42, 178–79
Wu Fa-hsien, 83

YAK-9P fighters (Yakovlev), 19, 20, 26, 209–10
YAK-11 trainers (Yakovlev), 20, 73, 210–11
YAK-12 communications aircraft (Yakovlev), 20, 42, 43, 73, 211–12
YAK-14 gliders (Yakovlev), 38, 212–13
YAK-16 trainers (Yakovlev), 20, 44, 213–14
YAK-17UTI trainers (Yakovlev), 24, 214–15
YAK-18 trainers (Shenyang), 174–75
YAK-18 trainers (Yakovlev), 34, 40, 73, 215–16
Yakovlev, Alexander S., 211
Yakovlev YAK-9P fighters, 19, 20, 26, 209–10
Yakovlev YAK-11 trainers, 20, 73, 210–11
Yakovlev YAK-12 communications aircraft, 20, 42, 43, 73, 211–12
Yakovlev YAK-14 gliders, 38, 212–13
Yakovlev YAK-16 trainers, 20, 44, 213–14
Yakovlev YAK-17UTI trainers, 24, 214–15
Yakovlev YAK-18 trainers, 34, 40, 73, 215–16
Yenan No. 1 transports, 43, 164–65

Zibin, P. V., 213